William Tebb, Edward Perry Vollum

Premature Burial and How it May be Prevented

With Special Reference to Trance, Catalepsy, and other Forms of Suspended

Animation

William Tebb, Edward Perry Vollum

Premature Burial and How it May be Prevented
With Special Reference to Trance, Catalepsy, and other Forms of Suspended Animation

ISBN/EAN: 9783337125370

Printed in Europe, USA, Canada, Australia, Japan

Cover: Foto ©berggeist007 / pixelio.de

More available books at **www.hansebooks.com**

PREMATURE BURIAL,

AND

HOW IT MAY BE PREVENTED.

PREMATURE BURIAL

AND

HOW IT MAY BE PREVENTED

*WITH SPECIAL REFERENCE TO TRANCE, CATALEPSY, AND
OTHER FORMS OF SUSPENDED ANIMATION*

BY

WILLIAM TEBB, F.R.G.S.

*Corresponding Member of the Royal Academy of Medical Sciences, Palermo;
Author of "The Recrudescence of Leprosy and its Causation"*

AND

Col. EDWARD PERRY VOLLUM, M.D.

*Late Medical Inspector, U.S. Army; Corresponding Member of the
New York Academy of Sciences*

LONDON
SWAN SONNENSCHEIN & CO., LIM.
1896

" What if in the tomb I awake !"—*Romeo and Juliet.*

" How comes it about that patients, given over as dead by their physicians, sometimes recover, and that some have even returned to life in the very time of their funerals?"—CELSUS.

" Such is the condition of humanity, and so uncertain is men's judgment, that they cannot determine even death itself."—PLINY.

PREFACE.

A DISTRESSING experience in the writer's family many years ago brought home to his mind the danger of premature burial, and led ultimately to the careful study of a gruesome subject to which he has a strong natural repugnance. His collaborator in the volume has himself passed through a state of profound suspended animation from drowning, having been laid out for dead—an experience which has induced him in like manner to investigate the various death-counterfeits. The results of the independent inquiries carried on by both of us in various parts of Europe and America, and by one of us during a sojourn in India in the early part of this year, are now laid before the reader, with such practical suggestions as it is hoped may prepare the way for bringing about certain needed reforms in our burial customs.

The danger, as I have attempted to show, is very real—to ourselves, to those most dear to us, and to the community in general ; and it should be a subject of very anxious concern how this danger may be minimised or altogether prevented. The duty of taking the most effective precautions to this end is

one that naturally falls to the Legislature, especially
under a Government professing to regard social questions
as of paramount importance. Fortunately, this is a non-
party and a non-contentious question, it imperils no
interest, so that no formal obstruction or unnecessary
delay need be apprehended ; and it should be urged
upon the Government to introduce and carry an effective
measure at the earliest opportunity, not only as a security
against the possibility of so terrible an evil, but to quiet
the widespread and not altogether unreasonable apprehension on this subject which is now so prevalent.

It has been found convenient to retain throughout the
body of the work the use of the singular pronoun, but
every part of the book receives the cordial approval of
both authors, and with this explanation we accept its
responsibility jointly.

We have to acknowledge our great indebtedness in
preparing this volume to many previous writers, including such as have investigated the phenomena of suspended animation and the signs of death, and such as,
with a more practical intention, have dwelt upon the
danger of death-counterfeits being mistaken for the
absolute extinction of life, illustrating their counsels or
warnings by numerous instances. Grouping both classes
of writers together, we may mention specially the names
of Winslow and Bruhier, Hufeland, Struve, Marcus Herz
and Köppen, Kite, Curry, and Anthony Fothergill ; and,
of more recent date, the names of Bouchut, Londe,

Lénormand, and Gaubert (on mortuaries). Russell
Fletcher, Franz Hartmann, and Sir Benjamin Ward
Richardson.

A work to which we are particularly indebted for the
literature of the subject is that of the late Dr. Félix
Gannal, "Mort Apparente et Mort Réelle : moyens de
les distinguer." Paris, 1890. Dr. Gannal, having qualified in medicine and pharmacy, occupied himself with
the business of embalming, which he inherited from his
father. He employed the considerable leisure which the
practice of that art left to him in compiling the above
laborious work. He examined many books, pamphlets,
theses, and articles, from which he cited expressions of
opinion on the several points—in a lengthy form in his
original edition (1868), in a condensed form in the second
edition. His Bibliography is by far the most comprehensive that has been hitherto compiled. Our own
Bibliography had been put together from various sources
before we made use of Dr. Gannal's. It includes several
titles which he does not give ; while, on the other
hand, it has been considerably extended beyond its
original limits by transcribing titles which we have found
nowhere but in his list. The Bibliography, it need
hardly be said, is much more extensive than our own
reading ; but it seemed useful to make it as complete as
possible, whether the books had been seen by us or not,
so as to show in chronological order how much interest
had been aroused in the subject from time to time—in

one country more than another, or in various countries
together.   The titles of articles in journals, which belong
for the most part to the more recent period, have been
taken from the Index Catalogue of the Surgeon-
General's Library, Washington, a few references being
added to articles which have otherwise come under our
notice.

W. T.

CONTENTS.

CHAPTER IX.

CHAPTER X.

CHAPTER XI.

CHAPTER XII.

CHAPTER XIII.

CHAPTER XIV.

CHAPTER XV.

CHAPTER XVI.

CHAPTER XVII.

CHAPTER XVIII.

CHAPTER XIX.

INTRODUCTION.

A CONCURRENCE of peculiar circumstances, beginning in May, 1895, has directed public attention in England to the subject of premature burial, probably to a greater degree, so far as the author's recollection serves, than at any time during the past half-century. Amongst these may be mentioned the publication of several recent cases of premature burial in the English and American papers ; the narrow escape of a child found in Regent's Park, London, laid out for dead at the Marylebone Mortuary, and afterwards restored to life ; the issue in Boston, U.S., of Dr. Franz Hartmann's instructive essay, entitled, " Buried Alive : an Examination into the Occult Causes of Apparent Death, Trance, and Catalepsy " (a considerable number of copies having been sold in England), and the able leading articles and correspondence on the subject in the *Spectator, Daily Chronicle, Morning Post, Leeds Mercury, The Jewish World, Plymouth Mercury, Manchester Courier, To-Day,* and many other daily and weekly journals.

It is curious, that while many books and pamphlets relating to this important subject have been issued in France and Germany, no adequate and comprehensive

treatise has appeared from the English press for more
than sixty years past, nor writings in any form, with
the exception of a paper by Sir Benjamin Ward
Richardson in No. 21 of the *Asclepiad*, published in
1889, on the "Absolute Signs of Death," sundry
articles in the medical journals from time to time,
and a London edition of Dr. Hartmann's volume in
January, 1896. The section upon "Real and Apparent
Death" in the 1868 edition of the late Professor Guy's
Forensic Medicine begins with the words, "This subject has never attracted much attention in England,
and no medical author of repute has treated it at any
length"—a remark not less true after the lapse of a
generation. The following chapters have been prepared with the view, not so much of supplying this
omission, as of guiding the public to the dangers of our
present mode of treating the apparent dead, in the hope
that reforms and preventive measures may be instituted
without delay in order to put an end to such unnecessary domestic tragedies.

In introducing the subject the author is aware that
the great majority of the medical profession in this
country are either sceptical or apathetic as to the
alleged danger of living burial. Many do not believe
in the existence of death-trance or death-counterfeits,
and the majority of those who do believe in them
declare that cases are very rare, and that if consciousness is ever restored in the grave it can only

last a second or two, and that those who live in fear
of such an occurrence should provide for a *post-
mortem* or for the severance of the jugular vein.
Many persons, on the other hand, after much careful
inquiry, are of opinion that cases of premature burial
are of frequent occurrence ; and that the great majority of the human race (outside of a few places in
Germany, where waiting mortuaries are established, or
where the police regulations, such as those described
in this volume as existing in Würtemburg, are
efficiently and systematically carried out) are liable to
this catastrophe. Important as the subject is allowed
to be, and numerous as are the reported cases, no
effective steps, either public or private, appear to have
been taken, outside of Germany and Austria, to remedy
the evil. At present a majority of the people appear
content to trust to the judgment of their relations
and to the ordinary certificates of death to safeguard
them from so terrible a disaster. That death-
certificates and death - verifications are often of a
most perfunctory description, both as to the fact of
death and the cause of death, has been proved by
overwhelming evidence before the recent House of
Commons Committee on Death - Certification. Such
certificates, when obtained, may be misleading and
untrustworthy ; while in many cases burials take place
without the doctor having either attended the patient or
examined the body. Nor, in spite of the appointment

of death-verificators by our neighbours across the
Channel, is this important precaution effectively carried
out by them. M. Devergie reports that in twenty-
five thousand communes in France no verification of
death takes place, although the law requires it; and
he demands that no diploma shall be given without
the candidate having proved himself conversant with
the signs of death. (*Medical Times*, London, 1874,
vol. i., p. 25.) On personal inquiry from medical
authorities in France, during the present year (1896),
we learn that this laxity still prevails.

It appears strange that, except when a man dies, all
his concerns are protected by custom and formalities, or
guarded by laws, so as to insure his interests being
fairly carried out to completion. Thus we see that
heirship, marriage, business affairs of all kinds, whether
of a public or private nature, are amply guarded by
such precautionary and authoritative measures as will
secure them. But one of the most important of all
human interests—that which relates to the termination
of life—is managed in such a careless and perfunctory
way as to permit of irreparable mistakes. To be sure
there are laws in most of the Continental States of
Europe that are intended to regulate the care and burial
of the dead, but few of them make it certain that the
apparently dead shall not be mistaken for the really dead,
and treated as such. None of them allow more than
seventy-two hours before burial (some allow only thirty-

six, others twenty-four, and others again much less, according to the nature of the disease), unless the attending physician petitions the authorities for reasonable delay—a rare occurrence. And even if postponement is granted, it is doubtful if the inevitable administrative formalities would leave opportunities for dubious cases to receive timely and necessary attention, or for cases of trance, catalepsy, coma, or the like, to be rescued from a living burial.

In the introduction to a Treatise entitled "The Uncertainty of the Signs of Death, and the Danger of Precipitate Interments," published in 1746, the author, Mr. M. Cooper, surgeon, says:—"Though death at some time or other is the necessary and unavoidable portion of human nature, yet it is not always certain that persons taken for dead are really and irretrievably deprived of life, since it is evident from experience that many apparently dead have afterwards proved themselves alive by rising from their shrouds, their coffins, and even from their graves. It is equally certain that some persons, too soon interred after their supposed decease, have in their graves fallen victims to a death which might otherwise have been prevented, but which they then find more cruel than that procured by the rope or the rack." The author quotes Lancisi, first physician to Pope Clement XI., who, in his Treatise *De subitaneis mortibus*, observes :—" Histories and relations are not the only proofs which convince me that many persons supposed to be dead have shown themselves alive, even when they were ready to be buried, since I am induced to such a belief from what I myself have seen ; for I saw a person of distinction, now alive, recover sensation and motion when the priest was performing the funeral service over him in church."

After reporting and describing a large number of cases of premature burial, or of narrow escapes from such terrible occurrences, in which the victims of hasty diagnosis were prepared for burial, or revived during the progress of the burial service, Mr. Cooper continues :—" Now, if a multiplicity of instances evince that many have the good fortune to escape being interred alive, it is justly to be suspected that a far greater number have fallen victims to a fatal confinement in their graves. But because human nature is such a slave to prejudice, and so tied down by the fetters of custom, it is highly difficult, if not absolutely impossible, to put people on their guard against such terrible accidents, or to persuade those vested with authority to take proper measures for preventing them."

Nothing seems to have been done to remedy this serious evil ; and forty-two years later Mr. Chas. Kite, a well-known practitioner, called attention to the subject in a volume, entitled " The Recovery of the Apparently Dead," London, 1788. This author, on p. 92, says :—" Many, various, and even opposite appearances have been supposed to indicate the total ex-

tinction of life. Formerly, a stoppage of the pulse
and respiration were thought to be unequivocal signs
of death; particular attention in examining the state
of the heart and larger arteries, the flame of a taper,
a lock of wool, or a mirror applied to the mouth or
nostrils, were conceived sufficient to ascertain these
points; *and great has been the number of those who
have fallen untimely victims to this erroneous opinion.*
Some have formed their prognostic from the livid,
black, and cadaverous countenance; others from the
heavy, dull, fixed, or flaccid state of the eyes; from
the dilated pupil; the foaming at the mouth and
nostrils, the rigid and inflexible state of the body,
jaws, or extremities; the intense and universal cold,
etc. Some, conceiving any one of these symptoms as
incompetent and inadequate to the purpose, have
required the presence of such of them as were, in
their opinion, the least liable to error; but whoever
will take the trouble of reading the Reports of the
(Humane) Society with attention, will meet with very
many instances where all the appearances separately,
and even where several associated in the same case,
occurred, and yet the patient recovered; and it is
therefore evident that these signs will not afford
certain and unexceptionable criteria by which we may
distinguish between life and death."

Mr. Kite furnishes references to numerous cases of
recovery where the apparently dead exhibited black,
livid, or cadaverous countenances; eyes fixed or

obscure; eyeballs diminished in size, immovable and fixed in their sockets, the cornea without lustre; eyes shrivelled; froth at the mouth; rigidity of the body, jaws, and extremities; partial or universal cold.[1]

The crux of the whole question is the uncertainty of the signs which announce the cessation of physical existence. Prizes have been offered, and prizes have been awarded, but further experience has shown that the signs and tests, sometimes singly and sometimes in combination, have been untrustworthy, and that the only certain and unfailing sign of death is decomposition.

Commenting upon actual cases of premature burial, the *Lancet*, March 17, 1866, p. 295, says:—"Truly there is something about the very notion of such a fate calculated to make one shudder, and to send a cold stream down one's spine. By such a catastrophe is not meant the sudden avalanche of earth, bricks, or stones upon the luckless miner or excavator, or the crushing, suffocative death from tumbling ruins. No; it is the cool, determined treatment of a living being as if he were dead—the rolling him in his winding sheet, the screwing him down in his coffin, the weeping at his funeral, and the final lowering of him into the narrow grave, and piling upon his dark and box-like dungeon loads of his mother earth. The last footfall departs

---

[1] "The Recovery of the Apparently Dead," by Charles Kite, Member of the Corporation of Surgeons in London, and Surgeon at Gravesend in Kent. London, 1788.

from the solitary churchyard, leaving the entranced
sleeper behind in his hideous shell soon to awaken to
consciousness and to a benumbed half-suffocated existence for a few minutes ; or else, more horrible still, there
he lies beneath the ground conscious of what has been
and still is, until, by some fearful agonised struggle of
the inner man at the weird phantasmagoria which has
passed across his mental vision, he awakes to a bodily
vivification as desperate in its torment for a brief period
as has been that of his physical activity. But it is
soon past. There is scarcely room to turn over in the
wooden chamber ; and what can avail a few shrieks and
struggles of a half-stifled, cramped-up man ! "

To prevent such unspeakable horrors as are here
pictured, the Egyptians kept the bodies of the dead under
careful supervision by the priests until satisfied that
life was extinct, previous to embalming them by means
of antiseptics, balsams, and odoriferous gums. The
Greeks were aware of the dangers of premature burial,
and cut off fingers before cremation to see whether
life was extinct. In ancient Rome the recurrence of
cases of premature burial had impressed the nation
with the necessity for exercising the greatest caution
in the treatment of the supposed dead ; hasty conclusions were looked upon as criminal, the absence of
breath or heat or a cadaverous appearance were
regarded as uncertain tests, and the supposed dead
were put into warm baths or washed with hot water,
and other means of restoration adopted. Neither in

the greater part of Europe nor in the United States
are any such means resorted to now, except in the
case of apparent death by drowning, by asphyxia, or
by hanging. Premature burials and narrow escapes are
of almost every-day occurrence, as the narratives in the
newspapers testify ; and the complaint made by a surgeon, Mr. Cooper, a hundred and fifty years ago, that the
evil is perpetuated because we are slaves to prejudice,
and because those vested with authority refuse to take
measures for prevention, remains a serious blot upon
our advanced civilisation. The *Spectator*, September
14, 1895, commenting upon this unsatisfactory state
of affairs, observes :—" Burning, drowning, even the
most hideous mutilation under a railway train, is as
nothing compared with burial alive. Strangely enough
this universal horror seems to have produced no desire
to guard against burial alive. We all fear it, and yet
practically no one takes any trouble to avoid the risk
of it happening in his own case, or in that of the rest
of mankind. It would be the simplest thing in the
world to take away all chance of burying alive ; and
yet the world remains indifferent, and enjoys its horror
undisturbed by the hope of remedy."

The authors' own reasonings, opinions, and conclusions
are here briefly presented; but as the majority of the
public are more or less influenced by authority, it has
been thought advisable to furnish a series of authenticated facts under the several headings to which they
belong, and to cite the judgments of eminent members

of the medical profession who have given special attention to the subject. The source of difficulty has been an *embarras de richesse*, or how from a mass of material, the extent of which will be seen by reference to the Bibliography, to select typical cases without needless repetition. The premature burials and narrow escapes from such disasters, which are reported by distinguished physicians and reputable writers, may be numbered literally by hundreds, and for every one reported it is obvious from the nature of the case that many are never heard of. Amongst the names of notable persons who have thought the subject sufficiently practical for their attention may be mentioned those of Empedocles, Plato, Aristotle, Cicero, Pliny, Celsus, Plutarch, and St. Augustine in antiquity; of Fabricius, Lancisi, Winslow, Haller, Buffon, Lavater, Moses Mendelssohn, Hufeland, and Alexander von Humboldt in modern times.

The subject has several times engaged the attention of the French Senate and Legislative Chamber, as well as the Legislative Assemblies in the various States of Germany. In 1871, Dr. Alex. Wilder, Prof. of Physiology and Psychological Science, read a paper before the members of both houses of the New York State Legislature at the Capitol, Albany; but we are not aware that the subject has ever been introduced in any of the other State Legislatures, or in the British Parliament, or in any of the Colonial Assemblies.

In an editorial note, as far back as November 27, 1858, the *Lancet*, referring to a case of death-trance, remarked that such "examples are sufficiently mysterious in their character to call for a more careful investigation than it has hitherto been possible to accord to them." The facts disclosed in this treatise, the authors hope, may encourage qualified scientific observers to study the subject of death-trance, which, it must be admitted, has been strangely overlooked in England, though it would not be easy to mention one which more deeply concerns every individual born into the world.

In order to prevent unnecessary pain to the reader on a subject so distressing in its nature, the more sensational and horrifying cases of premature burial have been omitted. They can, however, be found in abundance in the writings of Bruhier, Köppen, Kempner, Lénormand, Bouchut, Russell Fletcher, and the Boston (U.S.) edition of Hartmann. In England and in America it is the fashion amongst medical men to maintain that the tests known to medical art are fully equal to the prevention of live burial, that the cases quoted by the newspapers are introduced for sensational purposes, and that most of them are apocryphal. The perusal of the cases recorded in this volume, and a careful consideration of the weight of cumulative evidence represented by the very full bibliography, must satisfy the majority of reflective readers that the facts are both authentic and numerous.

PREMATURE BURIAL,

AND

HOW IT MAY BE PREVENTED.

SOME FORMS OF SUSPENDED ANIMATION.

CHAPTER I.

TRANCE.

OF all the various forms of suspended animation and apparent death, trance and catalepsy are the least understood, and most likely to lead the subject of them to a premature burial; the laws which control them have perplexed pathologists in all ages, and appear to be as insoluble as those which govern life itself. Dr. Le Clerc, in his "History of Medicine," records that " Heraclides, of Pontus, wrote a book *concerning the causes of diseases*, and another *concerning the disease in which the patient is without respiration*, in which he affirmed that in this disorder the patient sometimes continued thirty days without respiration, in such wise that he appeared dead, notwithstanding that there was no corruption of the body."[1]

---

[1] " Histoire de la Médecine," La Haye, 1729, p. 333.

Dr. Herbert Mayo, in "Letters on Truths Contained in Popular Superstitions," p. 34, says that "death-trance is the suspension of the action of the heart, and of breathing, and of voluntary motion—generally little sense of feeling and intelligence. With these phenomena is joined loss of external warmth, so that the usual evidence of life is gone. But there has occurred every shade of this condition that can be imagined, between occasional slight manifestations of suspension of one or other of the vital actions and their entire disparition."

Macnish, who also asserts that the function of the heart must go on, and even of the respiration, however slightly, says—"No affection to which the animal frame is subject is more remarkable than this (catalepsy, or trance). . . . There is such an apparent extinction of every faculty essential to life, that it is inconceivable how existence should go on during the continuance of the fit."—*Philos. of Sleep, Glasgow, 1834, pp. 225-6.*

In Quain's "Dictionary of Medicine," ii., p. 1063, Dr. Gowers says:—"The state now designated hypnotism is really induced trance, and trance has been accurately termed 'spontaneous hypnotism.' . . .

"The mental functions seem, in most cases, to be in complete abeyance. No manifestations of consciousness can be observed, or elicited by the most powerful cutaneous stimulation, and on recovery no recollection of the state is preserved. But in some cases volition only is lost, and the patient is aware of all that passes, although unable to give the slightest evidence of consciousness. . . .

"In the cases in which the depression of the vital

functions reaches an extreme degree, the patient appears
dead to casual and sometimes to careful observation.
This condition has been termed 'death-trance,' and
has furnished the theme for many sensational stories,
but the most ghastly incidents of fiction have been
paralleled by well-authenticated facts. [The last clause
appears in the new edition as follows:—" Persons have
certainly been buried in this state, and during the
recent epidemic of influenza an Italian narrowly escaped interment during the consequent trance."]

" The duration of trance has varied from a few hours
or days to several weeks, months, or even a year.

" Occasionally it is attended by some vaso - motor
disturbance. In a well-authenticated case of death-
trance the intense mental excitement produced by the
preparations for fastening the coffin lid occasioned a
sweat to break out over the body."

Many notable men have at one time or another been
subject to this disorder. Speaking of Benjamin Disraeli,
Mr. J. Fitzgerald Molloy, in his " Life of the Gorgeous
Lady Blessington,"vol.ii., pp. 37,38,says that in his "youth
he was seized with fits of giddiness, during which the world
swung round him, he became abstracted, and once fell
into a trance from which he did not recover for a week."

LETHARGIC STUPOR, OR TRANCE.

The *Lancet* of December 22, 1883, pp. 1078-80,
contains particulars from the pen of W. T. Gairdner,
M.D., LL.D., etc., Professor of Medicine in the University
of Glasgow, of a remarkable case of trance, extending
continuously over more than twenty-three weeks, which
attracted a considerable amount of notoriety at the time

and led to an extensive discussion. In his comments upon the case, the author continues, in the issue of January 5, 1884, pp. 5, 6 :—

"The case recorded in the *Lancet* of December 22, 1883, p. 1078, has been left up to this point without remarks, other than those obviously suggested by the direct observation of the facts in comparison or contrast with those of other cases coming more or less under the designation above mentioned. But in perusing, even in the most cursory manner, the multitudinous literature pertaining to the subjects of 'trance,' 'ecstasy,' 'catalepsy,' etc., not to speak of the popular narratives which from a very remote antiquity have handed down the tradition of preternatural sleep as an element in the fairy tales of almost all languages, one is struck by the almost uncontrollable disposition to regard such cases as altogether outside the limits of true physiological science : as being, according to the expressive Scotch phrase, 'no canny'—or, in other words, miraculous— and as involving questions connected with the unseen world, 'the undiscovered country from whose bourn no traveller returns.' So much is this the case, that, if in this nineteenth century the questions which presented themselves to Hippocrates in the treatise, περὶ ἱερῆς νούσου ('Concerning the Sacred Disease'), had to be rediscussed, it would certainly be in regard to some of the disorders mentioned above, and not as to epilepsy in its well-recognised clinical types, that the theory of a supernatural origin of the phenomena, whether favourably entertained or not, would fall to be argued. The irreconcilable differences of opinion in the Belgium Academy, as regards the quite modern instance of

Louise Lateau, are sufficient to show that all the culture and the scientific instincts of the present age have not quite inaugurated the 'reign of law,' nor established finally the position that 'miracles do not happen.' On the other hand, the researches of M. Charcot and others seem to be ever extending the domain of science further into the region of the marvellous and the obscure, so that even the most pronounced cases of 'demoniac possession' of the olden time have become the commonplaces of hystero-epilepsy in the clinique of the Salpêtrière. The peculiar interest of the present case is that it is altogether devoid of any of these adventitious, and more or less romantic, incidents. The patient is the mother of a family, and has lived a strictly domestic and (up to a short time before her seizure) healthy and regular life. There are no peculiar moral and religious problems to perplex the situation. There is no history of inveterate hysteria, or of long continued rapt contemplation ; nor has there been the slightest evidence of any craving after notoriety, either before the attack or since its termination. The moral atmosphere, in short, surrounding the phenomena is altogether unfavourable to exaggeration and imposture, for which, indeed, no reasonable motive can be assigned. Nevertheless, under these very commonplace conditions, concurring with some degree of melancholy or mental despondency after delivery, but during a convalescence otherwise normal, Mrs. M'I—— presents to our notice a condition of suspended consciousness and disordered innervation in no degree less extreme than the 'trances' or cataleptic attacks which have been recorded as the result of the most aggravated hysteria, or as the miracles of religious ecstasy and profound mental emotion. She becomes for the long period of over a hundred and sixty days continuously an almost mindless automaton, connected with the external world only through a few insignificant reflexes and through the organic functions. She is absolutely passive as regards everything that demands spontaneous movement, and betrays almost no sign of sensation, general or special, when subjected to the severest tests that can be applied short of physical injury."

In further notes upon the case, in the *Lancet* of January 12, 1884, p. 58, Professor Gairdner says :—

"The only other case to which I desire to make allusion at present is one in which I am, fortunately, in a position to furnish a sequel to an incomplete narrative, not without resemblance to the one lately published in this journal. 'A Case of Trance' was the subject of a paragraph in the *British Medical Journal* of May 31, 1879, p. 827, from which it appeared that in the London Hospital a woman, twenty-seven years of age, was at the time under the care of Dr. Langdon Down, being of rather small stature and weak mental capacity, and affected for at least two years with organic disease of the heart. About three weeks before the date of the report she had become suddenly somnolent, with most of the peculiarities in her sleep which have been already alluded to. She was fed partly by nutrient enemata, and for some days by a tube passed through the nostrils into the stomach. The resemblance is noted between this case and that of 'the famous Welsh fasting girl,' then attracting much attention in newspapers and otherwise. There being no further reference to this case in the journal, I wrote to Dr. Langdon Down, who kindly furnished me with the following additional particulars, which will, no doubt, be read even now with interest :—'My patient, who was in a state of trance, recovered somewhat suddenly after about four weeks, and left the hospital. The first indication of returning consciousness was observed when I was reading to my class at her bedside one of the numerous letters that I had received entreating me not to have her buried until something which the writers recommended had been done. The paragraph of the medical journal got into some Welsh paper, and then went the round of the provincial press, hence the number of letters I received. This special one was from an old gentleman of eighty-four years, who, when he was twenty-four, was thought to be dead, and whose friends had assembled to follow him to the grave, when he heard the undertaker say, "Would anyone like to see the corpse before I screw him down?" The undertaker at the same time moved the head a little and struck it against the coffin, on which he aroused and sat up. On reading this aloud a visible smile passed over the face of my patient, and she returned to obvious consciousness soon after. She has not come under observation since she left the hospital.'

"Although this case is probably only one among many, I mention it here because the receipt of the letter just given led me to investigate more particularly the state of the hearing in Mrs. M'I.'s case, and also to try the experiment of reading aloud Dr. Down's letter in her presence and that of the class. I had often remarked to bystanders that, although the subjects of these apparently unconscious states appeared inaccessible to the ordinary tests of sensibility, it was on record as regards some, even of those regarded as cases of 'apparent death,' that after recovery they affirm to have heard everything that passed, although unable to lift hand or foot to save themselves from premature burial. Neither the reading of the letter nor a violent shout into her ear produced any visible effects."

Thomas More Madden, M.D., F.R.C.S. (Edin.), in an article on "Death's Counterfeit," in the *Medical Press and Circular*, vol. i., April 27, 1887, pp. 386-8, relates the following case "of so-called hysteric trance" :—

"A young lady, Miss R——, apparently in perfect health, went to her room after luncheon to make some change in her dress.

A few minutes afterwards she was found lying on her bed in a profound sleep, from which she could not be awakened. When I first saw her, twenty-four hours later, she was sleeping tranquilly ; the decubitus being dorsal, respiration scarcely perceptible, pulse seventy, and extremely small; her face was pallid, lips motionless, and the extremities very cold. At this moment, so death-like was her aspect, that a casual observer might have doubted the possibility of the vital spark still lingering in that apparently inanimate frame, on which no external stimulus seemed to produce any sensorial impression, with the exception that the pupils were normal and responded to light. Sinapisms were applied over the heart and to the legs, where they were left on until vesication was occasioned without causing any evidence of pain. Faradisation was also resorted to without effect. In this state she remained from the evening of December 31 until the afternoon of January 3, when the pulse became completely imperceptible ; the surface of the body was icy cold, the respiratory movements apparently ceased, and her condition was to all outward appearance undistinguishable from death. Under the influence of repeated hypodermic injections of sulphuric ether and other remedies, however, she rallied somewhat, and her pulse and temperature improved. But she still slept on until the morning of the 9th, when she suddenly woke up, and, to the great astonishment of those about her, called for her clothes, which had been removed from their ordinary place, and wanted to come down to breakfast, without the least consciousness of what had occurred. Her recovery, I may add, was rapid and complete.

"The next case of lethargy that came under my notice was that of a boy, who, after an attack of fever, fell into a state of complete lethargic coma, in which he lay insensible between life and death for forty-seven days, and ultimately recovered perfectly.

"In a third instance of the same kind, in a lady under my care, the patient, after a lethargic sleep of twenty-seven days, recovered consciousness for a few hours, and then relapsed into her former comatose condition, in which she died.

"The fourth case of lethargy which I have seen was, like the first, a case of trance, which lasted for seventy hours, during which the flickering vital spark was only preserved from extinction by the involuntary action of the spinal and nervous centres. In this instance the patient finally recovered.

"The fifth and last instance of profound lethargy that has come within my own observation occurred last autumn in the Mater Misericordiæ Hospital in a young woman. . . . In that instance, despite all that medical skill could suggest or unremitting attention could do, it was found impossible to arouse the patient from the apparently hysterical lethargic sleep in which she ultimately sank and died."

I have referred to the foregoing cases, occurring in one physician's experience, as disproving the general opinion that lethargy or trance is so rarely met with as to be of little medical importance. For my own part, I have no doubt that these conditions are of far more frequent occurrence than is generally supposed. Moreover, I have had reason to know that death is occasionally so exactly thus counterfeited that there is good cause for fearing the probability of living interment in some cases of hasty burial.

Referring to death-trance, Dr. Madden observes, *ib.*, p. 388—"Death-trance, or that profound degree of lethargy which closely counterfeits death, deserves greater attention than is generally paid to it as a pathological condition, as well as a possible cause of premature interment. For, unless we reject every statement, however well authenticated, of those who have witnessed such cases, merely because their experience does not tally with our preconceived opinions and wishes, neither the frequent occurrence of death-trance nor the fearful results of its non-recognition can be questioned."

Mr. John Chippendale, F.R.C.S., writing to the *Lancet*, 1889, vol. i., p. 1173, on "Catalepsy. — Post-mortem Sweating," says :—

"I may mention that there is a record of a man who during an illness was seized with trance, though, as he lay in what Claudio calls 'cold abstraction,' he was aware of all that was passing.   At last, as he was about to be covered in his coffin, his mental condition was such that he broke into a profuse sweat, which was fortunately perceived, and he recovered and was able to recount his experiences."

It would appear from the following telegram through Reuter's Agency that trance is occasionally epidemic :—

[From *Daily Telegraph*, March 17, 1890.]

" A NEW DISEASE.

"Vienna, March 15, 1890.

" Several cases of a new disease, which originally appeared in Mantua immediately after the subsidence of the recent influenza epidemic, and to which the people of that city gave the name of 'La nonna'--*Anglice*, 'Falling asleep'—have occurred in the Comitat of Pressburg.

" Persons suffering from this complaint fall into a death-like trance, lasting about four days, out of which the patient wakes in a state of intense exhaustion.   Recovery is very slow, but, so far, no fatal case has been reported."

A correspondent writing to the *English Mechanic* September 13, 1895, says :—" I know one lady who has been three times prepared for burial, and very narrowly escaped it on the first occasion."   The author wrote to the writer for further details, and received a reply, dated September 19, 1895, from which it appears that the lady had married into a political family of considerable note, who would not care to have her identity disclosed.   My correspondent says :—" I know that she lay several days in a state not to be distinguished from death ; that she was in her coffin, and, I believe, showed signs of life just as the coffin was about to be closed. On two subsequent occasions she passed into similar trances ; but though believed to be dead, and treated as such, the previous experience prevented any idea of burial being entertained" until clear evidence of dissolution should appear.

The *New York Weekly Witness* of January 15, 1896, reports

"A LONG CATALEPTIC SLEEP.

"Information was received at Milford, Pa., last Friday, that William Depue, a prominent citizen of Bushkill, Pike County, whose mind for seven years has been a blank, had suddenly returned to consciousness.

"Seven years ago, while at work, Mr. Depue became ill. Doctors were summoned, but they could find no possible ailment. The sick man sank into a cataleptic sleep, from which medical science could not arouse him.

"At no time during the long period did he recognise any one, and food was given him through a tube inserted in his mouth. He lost no flesh, and was apparently as healthy as any man. Although the best medical men in the country were called to his bedside, his case baffled them all.

"Upon recovering his senses he set about his usual labours as if he had been asleep but the ordinary time. He remembers nothing that has taken place during his seven years' trance."

The following case appeared in the *Middlesbrough Daily Gazette*, February 9, 1896, and in a number of English papers :—

"The young Dutch maiden, Maria Cvetskens, who now lies asleep at Stevensworth, has beaten the record in the annals of somnolence. At the beginning of last month she had been asleep for nearly three hundred days. The doctors, who visit her in great numbers, are agreed that there is no deception in the case. Her parents are of excellent repute, and it has never occurred to them to make any financial profit out of the abnormal state of their daughter. As to the cause of the prolonged sleep, the doctors differ."

CHAPTER II.

CATALEPSY differs in some of its characteristics from trance, but the one is often mistaken for the other. It is not so much a disease as a symptom of certain nervous disorders, and to which women and children are more particularly liable. Catalepsy can be produced artificially by hypnotisation. Like trance, it has often been mistaken for death, and its subjects buried alive.

Dr. Franz Hartmann differentiates the two disorders as follows :—"There seems hardly any limit to the time during which a person may remain in a trance; but catalepsy is due to some obstruction in the organic mechanism of the body, on account of its exhausted nervous power. In the last case the activity of life begins again as soon as the impediment is removed, or the nervous energy has recuperated its strength."

Dr. Gowers, in Quain's "Dictionary of Medicine," ed. 1894, vol. i., pp. 284-5, describes catalepsy as belonging to both sexes, at all ages from six to sixty. It is a nervous affection, commonly associated with distinct evidence of hysteria, but said sometimes to occur as an early symptom of epilepsy. It is attended commonly with loss of consciousness. The limbs remain in the position they occupied at the onset, as if petrified. The whole or part of the muscles pass into a state of rigidity. In profound conditions sensibility is lost to touch, pain, and electricity ; and no reflex movements can be induced even by touching the conjunctiva, a state of mental trance being associated.

Cassell's Family Physician (by Physicians and Surgeons of the principal London Hospitals) describes this singular affection, as follows :—" Catalepsy is one of the strangest diseases possible. It is of rare occurrence, and some very sceptical people have even gone so far as to deny its existence. That is all nonsense, for catalepsy is just as much a reality as gout or bronchitis. A fit of catalepsy—for it is a paroxysmal disease—consists essentially in the sudden suspension of thought, feeling, and the power of moving. The patient remains in any position in which she—we say she, for it occurs mostly in women—happens to be at the moment of the seizure, and will, moreover, retain any posture in which she may be placed during the continuance of the fit. For example, you may stretch out the arms to their full length, and there they remain stretched out without showing the slightest tendency to drop. It does not matter how absurd or inconvenient or apparently fatiguing the position may be, it is maintained until altered by some one or until the fit is over. In these attacks there are no convulsions, but, on the contrary, the patient remains perfectly immobile. She is just like a waxen figure, or an inanimate statue, or a frozen corpse.

"Cataleptic fits vary very much, not only in their frequency, but in their duration. Sometimes they are very short indeed, lasting only a few minutes. In one case, that of a lady, they would sometimes come on when she was reading aloud. She would stop suddenly in the middle of a sentence, and a peculiar stiffness of the whole body would seize her, fixing the limbs immovably for several minutes. Then it would pass off, and the reading would be continued at the very word at which it had been interrupted, the patient being quite unconscious that anything had happened. But sometimes fits such as these may last for days and days together, and it seems not improbable that people may have been buried in this state in mistake for death."

The following case, contributed by Dr. Gooch, will further illustrate this malady:—

"A lady, who laboured habitually under melancholy, a few days after parturition was seized with catalepsy, and presented the following appearances:—She was lying in bed motionless and apparently senseless. It was thought the pupils of her eyes were dilated, and some apprehensions were entertained of effusion on the brain; but on examining them closely it was found they readily contracted when the light fell upon them. The only signs of life were warmth, and a pulse which was one hundred and twenty, and weak. In attempting to rouse her from this senseless state, the trunk of the body was lifted up and placed so far back as to form an obtuse angle with the lower extremities, and in this posture, with nothing to support her, she continued sitting for many minutes. One arm was now raised, and then the other, and in the posture they were placed they remained. It was a curious sight to see her sitting up staring lifelessly, her arms outstretched, yet without any visible signs of animation. She was very thin and pallid, and looked like a corpse that had been propped up and stiffened in that attitude. She was now taken out of bed and placed upright, and attempts were made to rouse her by calling loudly in her ears, but in vain; she stood up, indeed, but as inanimate as a statue. The slightest push put her off her balance, and she made no exertion to retain it, and would have fallen had she not been caught. She went into this state three times; the first lasted fourteen hours, the second twelve hours, and the third nine hours, with waking intervals of three days after the first fit, and of one day after the second ; after this time the disease assumed the ordinary form of melancholia.—*The Science and Practice of Medicine, by Sir W. Aitken, p. 357.*

Dr. John Jebb, F.R.S., cited in Reynolds' "System of Medicine," vol. ii., pp. 99-102, has recorded the following graphic case:—

" In the latter end of last year (*viz.*, 1781), I was desired to visit a young lady who, for nine months, had been afflicted with that singular disorder termed a catalepsy. Although she was prepared for my visit, she was seized with the disorder as soon as my arrival was announced. She was employed in netting, and was passing the needle through the mesh, in which position she immediately became rigid, exhibiting, in a very pleasing form, a figure of death-like sleep, beyond the power of art to imitate or the imagination to conceive. Her forehead was serene, her features perfectly composed. The paleness of her colour, her breathing at a distance being also scarcely perceptible, operated in rendering the similitude to marble more exact and striking. The positions of her fingers, hands, and arms were altered with difficulty, but they preserved every form of flexure they acquired : nor were the muscles of the neck exempted from this law, her head maintaining every situation in which the hand could place it as firmly as her limbs," etc.

Dr. King Chambers, after citing the above case in full, continues:—

" The most common exciting cause of catalepsy seems to be strong mental emotion. When Covent Garden Theatre was last burnt down, the blaze flashed in at the uncurtained windows of St. Mary's Hospital. One of my patients, a girl of twenty, recovering from low fever, was woke up by it, and exclaimed that the day of judgment was come. She remained in an excited state all night, and the next morning grew gradually stiff, like a

corpse, whispering (before she became quite insensible) that she
was dead. If her arm was raised, it remained extended in the
position in which it was placed for several minutes, and then
slowly subsided. The inelastic kind of way in which it retained
its position for a time, and then gradually yielded to the force of
gravity, reminded one more of a wax figure than of the marble
to which Dr. Jebb compares it. A strange effect was produced
by opening the eye-lid of one eye; the other eye remained closed,
and the raised lid after a time fell very slowly like the arm. A
better superficial representation of death it is difficult to conceive.
. . . In both these cases I convinced myself carefully that
there was no deception.

  "Other cases are of much longer duration. . . . The death-
like state may last for days. It may be mistaken for real death,
and treated as such. . . .

  "Any cases of apparent death that did occur (in former days)
were burnt, or buried, or otherwise put out of the way, and
were never more heard of. But after the establishment of
Christianity, tenderness, sometimes excessive, for the remains of
departed friends took the place of the hard, heathen selfishness.
The dead were kept closer to the congregations of the living,
as if to represent in material form the dogma of the Communion
of Saints. This led to the discovery that some persons, indeed
some persons of note (amongst others, Duns Scotus the theologian,
at Cologne), had got out of their coffins, and died in a vain
attempt to open the doors of their vaults."

  The author relates several other remarkable cases.
Here is one :—

  "I lighted accidently on another case, communicated to the
same scientific body (Acad. Royale des Sciences), by M. Imbert
in 1713. It is that of the driver of the Rouen diligence, aged
forty-five, who fell into a kind of soporific catalepsy on hearing
of the sudden death of a man he had quarrelled with. It appears
that 'M. Burette, under whose care he was at La Charité, made
use of the most powerful assistances of art—bleeding in the arms,
the foot, the neck, emetics, purgatives, blisters, leeches,' etc. At

last somebody 'threw him naked into cold water to surprise him.'
The effect surprised the doctors as much as the patient. It is
related with evident wonder how that 'he opened his eyes, looked
steadfastly, but did not speak.' His wife seems to have been a
prudent woman, for a week afterwards she 'carried him home,
where he is at present: they gave him no medicine; he speaks
sensibly enough, and mends every day.'"

The *Lancet*, 1870, vol. i., p. 1044, in its Paris correspondence says :—

"The following curious case is related as having occurred at
Dunkirk, on April 14, and as 'showing the utility of catalepsy.'
A young girl of seventeen years was seized with a violent attack of
epilepsy, and fell, on the above date, into a canal. A boatman
immediately jumped into the water to save her, and brought her to
the shore after twenty minutes. The most singular circumstance
connected with the accident is that, when the young girl was taken
out of the water, she presented all the symptoms of catalepsy.
Notwithstanding this long immersion, she was resuscitated, and
nothing afterwards transpired to cause any anxiety."

Mr. James Braid, M.R.C.S., in the *Medical Times*,
1850, vol. xxi., p. 402, narrates a case of a cataleptic
woman in the Manchester Royal Infirmary under the
care of Dr. John Mitchell, and writes:—

"Every variety of contrivance and torture was resorted to by
various parties who saw her, for the purpose of testing the degree
of her insensibility, and for determining whether she might not
be an impostor, but without eliciting the slightest indication of
activity of any of the senses ; . . . nevertheless she *heard and
understood all that was said and proposed to be done, and suffered
the most exquisite torture from various tests applied to her !.'*
A fact so important as this ought to be published in every journal
throughout the civilised world; so that in future professional men
might be thereby led to exercise greater discretion and mercy in
their modes of applying tests to such patients."

4

The *Somerset County Herald* (Taunton) of October 12
1895, has the following :—

"EXTRAORDINARY CASE OF TRANCE NEAR WEYMOUTH.

"The wedding nuptials of a sailor from H.M.S. *Alexandra* and
a young woman residing at Broadwey, who were recently married,
have been interrupted in a most unusual manner by the newly-
made bride falling into a trance.   On the day following the
wedding Mr. and Mrs. Mortimer, for such is the name of the
newly-espoused pair, went for a drive, and on returning in the
evening the bride, remarking that she did not feel very well, went
upstairs, and before long was in a sound sleep, which continued
throughout the night and far into the following day. The relatives
of the bride, remembering symptoms which she had previously
developed, then sent for Dr. Pridham, who at once pronounced
that the unfortunate young woman had fallen into a trance.   Dr.
Colmer, of Weymouth, was likewise called; but nothing that these
two medical gentlemen could do had the slightest effect in arousing their patient from the state of lethargy into which she had so
suddenly and unexpectedly relapsed.   In this condition she remained for a space of five days, when she gradually showed signs
of returning animation, and in the course of a few hours regained
consciousness, though she was then in a very exhausted condition.
After her awakening the young woman developed inflammation of
the legs, which was regarded as a very serious condition for her to
be in.   In an interview on Saturday, Dr. Pridham described the
trance as being exceedingly death-like in character, and added
that, in such trances as the one in question, in the past people
have no doubt been actually buried."

A report of this case appears in the *St. James's
Gazette.*

A less experienced practitioner would probably have
made out a death certificate, as in numerous similar
cases.

After burial we hear no more of them ; they may
have been buried in a death-like trance, but the medical

certificate, no matter how inconsiderately given, con-
signs them to perpetual silence beyond appeal or escape.
Family remonstrance is then unavailing, for, except
in cases of strong suspicion of poisoning, no Home
Secretary or Coroner would grant an order for ex-
humation.

The existence of trance, catalepsy, and other death
counterfeits, followed by hasty burial, has been alluded
to by reputable writers from time immemorial; and
while the veracity of these writers has remained un-
challenged, and their narratives are confirmed by hundreds of cases of modern experience, the effect on the
public mind has been only of a transitory character,
and nothing has been done either in England or
America to safeguard the people from such dreadful
mistakes.

CHAPTER III.

ANIMAL AND SO-CALLED HUMAN HIBERNATION.

THE following case of the jerboa, or jumping mouse, recorded last century by Major-General Thomas Davies, F.R.S., in the "Transactions of the Linnæan Society,"[1] will show how far a torpid mammal may be removed from the opportunity of breathing, and how imperceptibly, to the eyes of an observer, its torpid life passed into actual death :—

"With respect to the figure given of it in its dormant state (plate viii., fig. 6), I have to observe that the specimen was found by some workmen in digging the foundation for a summer house in a gentleman's garden, about two miles from Quebec, in the latter end of May, 1787. It was discovered enclosed in a ball of clay, about the size of a cricket ball, nearly an inch in thickness, perfectly smooth within, and about twenty inches under ground. The man who first discovered it, not knowing what it was, struck the ball with his spade, by which means it was broken to pieces, or the ball also would have been presented to me. The drawing will perfectly show how the animal is laid during its dormant state [a tawny mouse, with long hind legs and long tail, coiled up into a perfect ovoid, of which the two poles are the crown of the head and the rump.] How

---

[1] "Linnæan Transactions," 1797, vol. iv., p. 155. "An Account of the Jumping Mouse of Canada—*Dipus Canadensis.*"

long it had been under ground it is impossible to say ;
but as I never could observe these animals in any parts
of the country after the beginning of September, I conceive that they lay themselves up some time in that
month, or beginning of October, when the frost becomes
sharp ; nor did I ever see them again before the last
week of May, or beginning of June.   From their being
enveloped in balls of clay, without any appearance of
food, I conceive they sleep during the winter, and remain for that time without sustenance.   As soon as I
conveyed this specimen to my house, I deposited it, as
it was, in a small chip box, in some cotton, waiting with
great anxiety for its waking ; but that not taking place
at the season they generally appear, I kept it until I
found it began to smell : I then stuffed it, and preserved
it in its torpid position.   I am led to believe its not recovering from that state arose from the heat of my
room during the time it was in the box, a fire having
been constantly burning in the stove, and which in all
probability was too great for respiration.   .   .   ."

Mr. Braid, after citing facts as to higher animals, proceeds :—" There are other creatures which have not the
power of migrating from climes too intensely hot for the
normal exercise of their physical functions, and the lives
of these animals are preserved through a state of torpor
superinduced by the want of sufficient moisture, their
bodies being dried up from excessive heat.   This is the
case with snails, which are said to have been revived by
a little cold water being thrown on them, after having
remained in a dry and torpid state for fifteen years.
The *vibrio tritici* has also been restored, after perfect
torpidity and apparent death for five years and eight

months, by merely soaking it in water. Some small, microscopic animals have been apparently killed and revived again a dozen times by drying and then applying moisture to them. This is remarkably verified in the case of the wheel-animalcule. And Spallanzani states that some animalcules have been recovered by moisture after a torpor of twenty-seven years. According to Humboldt, again, some large animals are thrown into a similar state from want of moisture. Such he states to be the case with the alligator and boa-constrictor during the dry season in the plains of Venezuela, and with other animals elsewhere."—*On Trance and Human Hibernation, p. 47.*

Dr. Moore Russell Fletcher, in his treatise on "Suspended Animation," pp. 7, 8, observes:—"Snakes and toads live for a long time without air or food. The following experiment was made by a Mr. Tower, of Gardiner (Maine). An adder, upwards of two feet in length, was got into a glass jar, which was tightly sealed. He was kept there for sixteen months without any apparent change, and when let out, looked as well as when put in, and crawled away.

"The common pond trout, when thrown into snow, will soon freeze, remain so for days, and when put into cold water to remove the frost become lively as ever.

"When residing in New Brunswick, in 1842, we went to a lake to secure some trout, which were frozen in the snow and kept for use. While there we saw men with long wooden tongs catching frost fish from the salt water at the entrance of a brook. The fish were thrown upon the ice in great quantities. We had a barrel of them put up with snow and kept frozen, and in a cool place. For six or seven weeks they were taken out and used as wanted, and might be kept frozen for an indefinite time, and be alive when thawed in cold water. The two pieces of a fish, cut in two when frozen, would move and try to swim when thawed in cold water."

SO-CALLED HUMAN HIBERNATION.

Dr. George Moore observes that "A state of the body is certainly sometimes produced (in man) which is nearly analogous to the torpor of the lower animals— *a condition utterly inexplicable to any principle taught in the schools.* Who, for instance, can inform us how it happens that certain fishes may be suddenly frozen in the Polar Sea, and so remain during the long winter and yet be requickened into full activity by returning summer?"—*Use of the Body in Relation to the Mind, p. 31.*

Hufeland, in his " Uncertainty of Death," 1824, p. 12, observes that it is easier for mankind to fall into a state of trance than the lower creatures, on account of their complicated anatomy. It is a transitory state between life and death, into which anyone may pass and return from. Trance was common among the Greeks and Romans, who, just before cremation, had the custom of cutting off a finger-joint, most probably to discover if there was any trace of life. Death does not come suddenly ; it is a gradual process from actual life into apparent death, and from that to actual death. It is a mistake to take outward appearances for inner death.

" It often happens a person is buried in a trance knowing all the preparations for the interment, and this affects him so much that it prolongs the trance by its

depressing influence. How long can a man exist in a
state of trance? Is there no sign by which the remaining spark of life may be recognised? Do no means
exist to prevent awakening in the grave? Nothing can
be said as to its duration; but we do know that differences in the cause and circumstances will cause a difference in duration. The amount of strength of the person
would have great effect in this. Weak persons, broken
down by excesses, would die sooner than the strong.
The nature of the disease would make a difference.
Old age is less liable to trance than the young. Long
sickness destroys the sources of life, and shortens the
process of death. Sorrow and trouble, and numerous
diseases, seem to bring on death; yet ofttimes the source
of life in them exists to its full extent, and what seems
in them to be death may be only a fainting fit, or cramp,
which temporarily interrupts the action of life. Women
are more liable to trance than men: most cases have
happened in them. Trance may exist in the new-born;
give them time, and many of them revive. The smell
of the earth is at times sufficient to wake up a case of
trance. Six or seven days, or longer, are often required
to restore such cases." (Extracted from pp. 10-24.)

Mr. Chunder Sen, municipal secretary to the Maha-
rajah of Jeypore, introduced the author, during his visit
to India, March 8, 1896, to a venerable and learned
fakir, who was seated on a couch Buddhist fashion, the
feet turned towards the stomach, in the attitude of
meditation, in a small but comfortable house near the
entrance to the beautiful public gardens of that city.
The fakir possesses the power of self-induced trance,
which really amounts to a suspension of life, being

indistinguishable from death. In the month of December, 1895, he passed into and remained in this condition for twenty days. On several occasions the experiment has been conducted under test conditions. In 1889, Dr. Hem Chunder Sen, of Delhi, and his brother, Mr. Chunder Sen, had the opportunity of examining the fakir while passing into a state of hibernation, and found that the pulse beat slower and slower until it ceased to beat at all. The stethoscope was applied to the heart by the doctor, who failed to detect the slightest motion. The fakir, covered with a white shroud, was placed in a small subterraneous cell built of masonry, measuring about six feet by six feet, of rotund structure. The door was closed and locked, and the lock sealed with Dr. Sen's private seal and with that of Mr. Dhanna Tal, the magistrate of the city; the flap door leading to the vault was also carefully fastened. At the expiration of thirty-three days the cell was opened, and the fakir was found just where he was placed, but with a death-like appearance, the limbs having become stiff as in *rigor mortis.* He was brought from the vault, and the mouth was rubbed with honey and milk, and the body and joints massaged with oil. In the evening, manifestations of life were exhibited, and the fakir was fed with a spoonful of milk. The next day he was given a little juice of pulses known as *dal,* and in three days he was able to eat bread and milk, his normal diet. These cases are well known both at Delhi and at Jeypore, and the facts have never been disputed. The fakir is a Sanscrit scholar, and is said to be endowed with much wisdom, and is consulted by those who are interested in Hindu learning and religion. He has never received

money from visitors, and the mention of it distresses
him.

The *Medical Times* of May 11, 1850, contains a communication from Mr. Braid, who says he has "lost no
opportunity of accumulating evidence on this subject,
and that while many alleged feats of this kind are
probably of a deceptive character, still there are others
which admit of no such explanation ; and that it becomes the duty of scientific men fairly to admit the
difficulty." He then refers to two documents by eye-
witnesses of these feats, and which, he says, "with the
previous evidence on the subject, must set the point at
rest for ever, as to the fact of the feats referred to being
genuine phenomena, deception being impossible." In
one of these instances, the fakir was buried in the
ground for six weeks, and was, consequently, deprived
not only of food and drink, but also of light and air;
when he was disinterred, his legs and arms were shrivelled and stiff, but his face was full ; no pulse could
be discovered in the heart, temples, or arms. "About
three years since I spent some time with a General
C——, a highly respectable and intelligent man, who
had been a long time in the Indian service, and who
was himself an eye-witness of one of these feats. A
fakir was buried several feet in the earth, under vigilant
inspection, and a watch was set, so that no one could
communicate with him; and to make the matter doubly
sure, corn was sown upon the grave, and during the
time the man was buried, it vegetated and grew to the
height of several inches. He lay there forty-two days.
The gentleman referred to passed the place many times
during his burial, saw the growing corn, was also present

at his disinterment, and when he questioned the man, and intimated to him that he thought deception had been practised, the fakir offered, for a sum of money, to be buried again, for the same length of time, by the General himself, and in his own garden. This challenge, of course, closed the argument."

Cases of this kind might be multiplied on evidence which cannot be doubted, and, in Mr. Braid's book, entitled "Human Hibernation," there are cases fully stated. Sir Claude Wade, who was an eye-witness of these feats when acting as political agent at the Court of Runjeet Singh, at Lahore, and from whom Mr. Braid derived his information, makes the following observations :—" I share entirely in the apparent incredibility of the fact of a man being buried alive and surviving the trial for various periods of duration ; but however incompatible with our knowledge of physiology, in the absence of any visible proof to the contrary, I am bound to declare my belief in the facts which I have represented, however impossible their existence may appear to others." Upon this Mr. Braid observes :—" Such then is the narrative of Sir C. M. Wade, and when we consider the high character of the author as a gentleman of honour, talents, and attainments of the highest order, and the searching, painstaking efforts displayed by him throughout the whole investigation, and his close proximity to the body of the fakir, and opportunity of observing minutely every point for himself, as well as the facilities, by his personal intercourse with Runjeet Singh and the whole of his Court, of gaining the most accurate information on every point, I conceive it is impossible to have had a more valuable or conclusive document for determining the fact that no collusion or deception existed."

A case of this kind was exhibited at the Westminster Aquarium in the autumn of 1895, which was carefully watched and tested by medical experts, without detection of any appearance of fraud or simulation. The hypnotised man, Walter Johnson, an ex-soldier, twenty-nine years of age, was in a trance which lasted thirty days, during which time he was absolutely unconscious, as shown by the various experiments to which he was subjected.

A case of induced trance and experimental burial, not unlike that of the Indian fakirs referred to, was reported in the London *Daily Chronicle*, March 14, 1896. The experiment was carried out under test conditions.

"'BURIED ALIVE' AT THE ROYAL AQUARIUM.

"After being entombed for six days in a hypnotic trance, Alfred Wootton was dug up and awakened at the Royal Aquarium (Westminster), on Saturday night in the presence of a crowd of interested spectators. Wootton was hypnotised on Monday by Professor Fricker, and consigned to his voluntary grave, nine feet deep, in view of the audience, who sealed the stout casket or coffin in which the subject was immured. Seven or eight feet of earth were then shovelled upon the body, a shaft being left open for the necessary respiration, and in order that the public might be able to see the man's face during the week. The experiment was a novel one in this country, and was intended to illustrate the extraordinary effect produced by the Indian fakirs, and to demonstrate the connection between hypnotism and psychology, while also showing the value of the former art as a curative agent. Wootton is a man thirty-eight years of age; he is a lead-worker, and on Monday weighed 10st. 2½lbs. He had previously been in a trance for a week in Glasgow, under Professor Fricker's experienced hands, so was not altogether new to the business; but he is

the first to be 'buried alive' by way of amusement. To the un-
initiated the whole thing was gruesome in the extreme, and this
particular form of entertainment certainly cannot be commended.
Before being covered in, Wootton's nose and ears were stopped
with wax, which was removed before he was revived on Saturday.
The theory of the burial is to secure an equable temperature day
and night—which is impossible when the subject is above ground
in the ordinary way—and therefore to induce a deeper trance. Of
course, too, the patient was out of reach of the operator, and no
suspicion of continuous hypnotising could rest upon the professor.
No nourishment could be supplied for the same reason, though
the man's lips were occasionally moistened by means of a damp
sponge on the end of a rod, and no record of temperature or
respiration could be kept. A good many people witnessed the
digging up process, and the awakening took place in the concert
room, whither the casket and its burden were conveyed. The
professor was not long in arousing his subject, after electric and
other tests had been applied to convince the audience that the
man was perfectly insensible to pain and everything else. Indeed,
a large needle was run through the flesh on the back of the hand
without any effect whatever. The first thing on regaining consciousness that Wootton said was that he could not see, and then
he asked for drink—milk, and subsequently a little brandy, being
supplied. As soon as possible the patient was lifted out of his box,
and with help was quickly able to walk about the platform. He
complained of considerable stiffness of the limbs, and was undoubtedly weak, but otherwise seemed none the worse for his
remarkable retirement from active life, and abstention from food
for nearly a week. He was swathed in flannel, and soon found the
heat of the room very oppressive, though at first he appeared to
be particularly anxious to have his overcoat and his boots. It
is anticipated that in a day or two at most Wootton will have
regained his usual vigorous health."

Dr. Hartmann in " Premature Burial," page 23, relates an account of a similar experiment with a fakir,
differing from the above, however, in so far as it was

made by some English residents, who did not put the
coffin into the earth, but hung it up in the air, so as to
protect it from the danger of being eaten up by white
ants. There seems to be hardly any limitation in regard
to the time during which such a body may be preserved
and become reanimated again, provided that it is well
protected, although modern ignorance may smile at this
statement.

Those of our readers who wish to pursue this subject
will find ample material in " Observations on Trance or
Human Hibernation," 1850, by James Braid, M.R.C.S.;
Dr. Kuhn's report of his investigations of the Indian
fakirs to the Anthropological Society of Munich, in
1895 ; the researches of Dr. J. M. Honigberger, a
German physician long resident in India ; and in the
India Journal of Medical and Physical Science, 1836,
vol. i., p. 389, etc.

CHAPTER IV.

PREMATURE BURIAL.

AT the sitting of the Paris Academy of Medicine, on April 10, 1827, a paper was read by M. Chantourelle, on the danger of hasty burial. This led to a discussion, in which M. Desgenettes stated that he had been told by Dr. Thouret, who presided at the destruction of the vaults of Les Innocens, that many skeletons had been found in positions seeming to show that they had turned in their coffins. Dr. Thouret was so much impressed by the circumstance that he had a special clause inserted in his will relating to his own burial.[1]

Similar revelations, according to Kempner, have followed the examinations of graveyards in Holland, and in New York and other parts of the United States.

On July 2, 1896, the author visited the grave of Madam Blunden, in the Cemetery, Basingstoke, Hants, who, according to the inscription (now obliterated), was buried alive. The following narrative appears in "The Uncertainty of the Signs of Death," by Surgeon M. Cooper, London, 1746, pp. 78, 79 :—

"At Basingstoke, in Hampshire, not many years ago, a gentlewoman of character and fortune was taken ill, and, to all appearance, died, while her husband was on a journey to London. A messenger was forthwith despatched to the gentleman, who returned immediately,

---

[1] Archives gén de Med., 1827, xiv., p. 105.

and ordered everything for her decent interment. Accordingly, on the third day after her supposed decease, she was buried in Holy Ghost Chapel, at the outside of the town, in a vault belonging to the family, over which there is a school for poor children. endowed by a charitable gentleman in the reign of Edward VI. It happened the next day that the boys, while they were at play, heard a noise in the vault, and one of them ran and told his master, who, not crediting what he said, gave him a box on the ear and sent him about his business; but, upon the other boys coming with the same story, his curiosity was awakened, so that he sent immediately for the sexton, and opened the vault and the lady's coffin, where they found her just expiring. All possible means were used to recover her to life, but to no purpose, for she, in her agony, had bit the nails off her fingers, and tore her face and head to that degree, that, notwithstanding all the care that was taken of her, she died in a few hours in inexpressible torment."

The *Sunday Times*, London, December 30, 1838, contains the following:—

"A frightful case of premature interment occurred not long since, at Tonneins, in the Lower Garonne. The victim, a man in the prime of life, had only a few shovelfuls of earth thrown into his grave, when an indistinct noise was heard to proceed from his coffin. The grave-digger, terrified beyond description, instantly fled to seek assistance, and some time elapsed before his return, when the crowd, which had by this time collected in considerable numbers round the grave, insisted on the coffin being opened. As soon as the first boards had been removed, it was ascertained, beyond a doubt, that the occupant had been interred alive. His countenance was frightfully contracted with the agony he had undergone; and, in his struggles, the unhappy man had forced his arms completely out of the winding sheet, in which they had been securely enveloped. A physician, who was on the spot, opened a vein, but no blood followed. The sufferer was beyond the reach of art."

Mr. Oscar F. Shaw, Attorney-at-Law, 145 Broadway, New York, furnished the author with particulars of the following case, of which he had personal knowledge :—
"In or about the year 1851, Virginia M'Donald, who, up to that time had lived with her father on Catharine Street, in the City of New York, apparently died, and was buried in Greenwood Cemetery, Brooklyn, N.Y.

"After the burial her mother declared her belief that the daughter was not dead when buried, and persistently asserted her belief. The family tried in various ways to assure the mother of the death of her daughter, and even resorted to ridicule for that purpose ; but the mother insisted so long and so strenuously that her daughter was buried alive, that finally the family consented to having the body taken up, when to their horror, they discovered the body lying on the side, the hands badly bitten, and every indication of a premature burial."

The *Lancet*, May 22, 1858, p. 519, has the following :—

"INTERMENT BEFORE DEATH.

"A case of restoration to consciousness after burial is recorded by the Austrian journals in the person of a rich manufacturer, named Oppelt, at Rudenberg. He was buried fifteen years ago, and lately, on opening the vault, the lid of the coffin was found forced open, and his skeleton in a sitting posture in a corner of the vault. A Government Commission has reported on the matter."

5

From the *Lancet*, August 20, 1864, p. 219.

"PREMATURE INTERMENT.

"Amongst the papers left by the great Meyerbeer, were some which showed that he had a profound dread of premature interment. He directed, it is stated, that his body should be left for ten days undisturbed, with the face uncovered, and watched night and day. Bells were to be fastened to his feet. And at the end of the second day veins were to be opened in the arm and leg. This is the gossip of the capital in which he died. The first impression is that such a fear is morbid. No doubt fewer precautions would suffice, but now and again cases occur which seem to warrant such a feeling, and to show that want of caution may lead to premature interment in cases unknown. An instance is mentioned by the *Ost. Deutscher Post* of Vienna. A few days since, runs the story, in the establishment of the Brothers of Charity in that capital, the bell of the dead-room was heard to ring violently, and on one of the attendants proceeding to the place to ascertain the cause, he was surprised at seeing one of the supposed dead men pulling the bell-rope. He was removed immediately to another room, and hopes are entertained of his recovery."

From the *Times*, July 7, 1867, p. 12, col. 3.

"The *Journal de Pontarlier* relates a case of premature interment. During the funeral, three days back, of a young woman at Montflorin, who had apparently died in an epileptic fit, the grave-digger, after having thrown a spadeful of earth on the coffin, thought he heard a moaning from the tomb. The body was consequently exhumed, and a vein having been opened, yielded blood almost warm and liquid. Hopes were for a moment entertained that the young woman would recover from her lethargy, but she never did so entirely, and the next day life was found to be extinct."

From the *Lancet*, October 19, 1867, p. 504.

"BURIED ALIVE.

"The *Journal de Morlaix* mentions that a young woman at Bohaste, France, who was supposed to have died from cholera a few days back, was buried on the following afternoon. The

sexton, when about to fill in the grave, fancied that he heard a
noise in the coffin, and sent for the medical officer, who, on removing the lid and examining the body, gave it as his opinion
that the woman had been alive when buried."

The official journal of the French Senate, January 30,
1869, records that the attention of the Senate was called
to this case by means of a petition signed by seven
residents in Paris, and the facts are confirmed by L.
Roger, *Officier de Santé*.

From the *Times*, May 6, 1874, p. 11, foot of col. 4.

" PREMATURE INTERMENT.

"The *Messager du Midi* relates the following dreadful story :—
A young married woman residing at Salon (Bouches du Rhône)
died shortly after her confinement in August last. The medical
man, who was hastily summoned when her illness assumed a
dangerous form, certified her death, and recommended immediate
burial in consequence of the intense heat then prevailing, and
six hours afterwards the body was interred. A few days since,
the husband having resolved to re-marry, the mother of his late
wife desired to have her daughter's remains removed to her native
town, Marseilles. When the vault was opened a horrible sight
presented itself. The corpse lay in the middle of the vault, with
dishevelled hair and the linen torn to pieces. It evidently had
been gnawed in her agony by the unfortunate victim. The shock
which the dreadful spectacle caused to the mother has been so
great that fears are entertained for her reason, if not for her life."

The *British Medical Journal*, December 8, 1877,
p. 819, inserts the following :—

"BURIED ALIVE.

"A correspondent at Naples states that the Appeal Court has
had before it a case not likely to inspire confidence in the minds
of those who look forward with horror to the possibility of being
buried alive. It appeared from the evidence that some time ago
a woman was interred with all the usual formalities, it being

believed that she was dead, while she was only in a trance. Some days afterwards, the grave in which she had been placed being opened for the reception of another body, it was found that the clothes which covered the unfortunate woman were torn to pieces, and that she had even broken her limbs in attempting to extricate herself from the living tomb. The Court, after hearing the case, sentenced the doctor who had signed the certificate of decease, and the mayor who had authorised the interment, each to three months' imprisonment for involuntary manslaughter."

From the *Daily Telegraph*, January 18, 1889.

"A gendarme was buried alive the other day in a village near Grenoble. The man had become intoxicated on potato brandy, and fell into a profound sleep. After twenty hours passed in slumber, his friends considered him to be dead, particularly as his body assumed the usual rigidity of a corpse. When the sexton, however, was lowering the remains of the ill-fated gendarme into the grave, he heard moans and knocks proceeding from the interior of the 'four-boards.' He immediately bored holes in the sides of the coffin, to let in air, and then knocked off the lid. The gendarme had, however, ceased to live, having horribly mutilated his head in his frantic but futile efforts to burst his coffin open."

The *Undertakers' and Funeral Directors' Journal*, July 22, 1889, relates the following cases :—

"A New York undertaker recently told the following story, the circumstances of which are still remembered by old residents of the city :—'About forty years ago a lady living on Division Street, New York City, fell dead, apparently, while in the act of dancing at a ball. It was a fashionable affair, and being able to afford it, she wore costly jewellery. Her husband, a flour merchant, who loved her devotedly, resolved that she should be interred in her ball dress, diamonds, pearls, and all ; also that there should be no autopsy. As the weather was very inclement when the funeral reached the cemetery, the body was placed in the receiving vault for burial next day. The undertaker was not a poor man, but he was avaricious, and he made up his mind to possess the jewellery. He went in the night, and took the lady's watch from the folds of her dress. He next began to draw a diamond ring from her finger, and in doing so had to use violence enough to tear the skin. Then the lady moved and groaned, and the thief, terrified and conscience-stricken, fled from the cemetery, and has never been since heard from, that I know of. The lady, after the first emotions of horror at her unheard-of position had passed over, gathered her nerves together and stepped out of the vault, which the thief had left open. How she came home I cannot tell; but this I know—she lived and had children, two at least of whom are alive to-day.'

"Another New York undertaker told this story. The New York papers thirty-five years ago were full of its ghastly details. 'The daughter of a Court Street baker died. It was in winter, and the father, knowing that a married sister of his dead child, who lived in St. Louis, would like to see her face before laid in the grave for ever, had the body placed in the vault, waiting her arrival. The sister came, the vault was opened, the lid of the coffin taken off, when, to the unutterable horror of the friends assembled, they found the grave-clothes torn in shreds, and the fingers of both hands eaten off. The girl had been buried alive.'

"Until about forty years ago a noted family of Virginia preserved a curious custom, which had been religiously observed for more than a century. Over a hundred years ago a member of the family died, and, upon being exhumed, was found to have been buried alive. From that time until about 1850, every member of the family, man, woman, or child, who died, was stabbed in the heart with a knife in the hands of the head of the house. The reason for the cessation of this custom was that in 1850 or thereabouts a beautiful young girl was supposed to be dead, the knife was plunged into her bosom, when she gave vent to a fearful scream and died. She had merely been in a trance. The incident broke her father's heart, and in a fit of remorse he killed himself not long afterwards.

"There are many families in the United States who, when any of their number dies, insist that an artery be opened to determine whether life has fled or not."

The following remarkable case of waking in the grave
is reported from Vienna :—

"A lady residing at Derbisch, near Kolin, in Bohemia, where
she owned considerable property, was buried last week, after a
brief illness, in the family vault at the local cemetery. Four days
afterwards her granddaughter was interred in the same place, but
as the stone slab covering the aperture was removed, the bystanders were horrified to see that the lid of the coffin below had
been raised, and that the arm of the corpse was protruding. It
was ascertained eventually that the unfortunate lady, who was
supposed to have died of heart disease, had been buried alive.
She had evidently recovered consciousness for a few minutes, and
had found strength enough to burst open her coffin. The authorities are bent on taking measures of the utmost severity against
those responsible."—*Undertakers' Journal, August 22, 1889.*

The *Undertakers' and Funeral Directors' Journal,*
July 22, 1890.

"A horrible story comes from Majola, Mantua. The body of
a woman, named Lavrinia Merli, a peasant, who was supposed to
have died from hysterics, was placed in a vault on Thursday,
July 3. On Saturday evening it was found that the woman had
regained consciousness, torn her grave-clothes in her struggles,
had turned completely over in the coffin, and had given birth to a
seven-months'-old child. Both mother and child were dead when
the coffin was opened for the last time previous to interment."

"A shocking occurrence is reported from Cesa, a little village
near Naples. A woman living at that place was recently seized
with sudden illness. A doctor who was called certified that the
woman was dead, and the body was consequently placed in a
coffin, which was deposited in the watch-house of the local
cemetery. Next day an old woman passing close to the cemetery
thought she heard smothered cries proceeding from the watch-
house. The family was informed, but when the lid of the coffin
was forced off a shocking spectacle presented itself to the gaze of
the horrified villagers. The wretched woman had turned on her

side, and the position of her arm showed that she had made
a desperate effort to raise the lid. The eldest son, who was
among the persons who broke open the coffin, received such
a shock that he died three days later."—*Undertakers' Journal*,
September 22, 1893.

The *Progressive Thinker*, of November 14, 1891, relates that :—

"Farmer George Hefdecker, who lived at Erie, Pa., died very
suddenly two weeks ago, of what is supposed to have been heart
failure. The body was buried temporarily four days later in a
neighbour's lot in the Erie cemetery pending the purchase of one
by his family. The transfer was made in a few days, and when
the casket was opened at the request of his family, a horrifying
spectacle was presented. The body had turned round, and the
face and interior of the casket bore the traces of a terrible struggle
with death in its most awful shape. The distorted and blood-
covered features bore evidence of the agony endured. The clothing about the head and neck had been torn into shreds, as was
likewise the lining of the coffin. Bloody marks of finger nails
on the face, throat, and neck, told of the awful despair of the
doomed man, who tore his own flesh in his terrible anguish.
Several fingers had been entirely bitten off, and the hands torn
with the teeth until they scarcely resembled those of a human
being."

From the London *Echo*, October 6, 1894.

"BURIED ALIVE.

"A story of a horrible nature comes from St. Petersburg in
connection with the interment at Tioobayn, near that city, of a
peasant girl named Antonova. She had presumably died, and in
due course the funeral took place. After the service at the
cemetery, the grave-diggers were startled by sounds of moaning
proceeding from the coffin. Instead, however, of instantly break-
ing it open, they rushed off to find a doctor, and when he and some
officials arrived and broke open the shell, the unhappy inmate was
already the corpse she had been supposed to be a day earlier. It

was evident, however, that no efforts could have saved life at the last moment. The body was half-turned in the coffin, the left hand, having escaped its bandages, being under the cheek."

The following case, cabled by Dalziel, appears in the London *Star*, August 19, 1895 :—

"SOUNDS FROM ANOTHER COFFIN.

"Grenoble, August 17.

"On Monday last a man was found in a dying condition by the side of a brook near the village of Le Pin. Everything possible was done for him, but he relapsed into unconsciousness, and became to all appearances dead. The funeral was arranged, and, there being no suspicion of foul play, the body was interred on the following day. The coffin had been lowered to the bottom of the grave, and the sexton had begun to cover it with earth, when he heard muffled sounds proceeding from it. The earth was hastily removed and the coffin opened, when it was discovered that the unfortunate occupant was alive. He was taken to a neighbouring house, but rapidly sank into a comatose condition, and died without uttering a word. The second burial took place yesterday."

While in India, in the early part of this year (1896), Dr. Roger S. Chew, of Calcutta, who, having been laid out for dead, and narrowly escaped living sepulture, has had the best reasons for studying the subject, gave me particulars of the following cases:—

"Frank Lascelles, aged thirty-two years, was seated at breakfast with a number of us young fellows, and was in the middle of a burst of hearty laughter when his head fell forward on his plate and he was 'dead.' As there was a distinct history of cardiac disease in his family, while he himself had frequently been treated for valvular disease of the heart, he was alleged to have 'died' of cardiac failure, and was duly interred in the Coonor Cemetery. Some six months later, permission was obtained to remove his remains to St. John's Churchyard in Ootacamund. The coffin was ex-

humed, and, as a 'matter of form,' the lid removed to identify the
resident, when, to the horror of the lookers-on, it was noticed that,
though mummification had taken place, there had been a fearful
struggle underground, for the body, instead of being on its back as
it was when first coffined, was *lying on its face,*with its arms and
legs drawn up as close as the confined space would permit.   His
trousers (a perfectly new pair) were burst at the left knee, while
his shirt-front was torn to ribands and bloodstained, and the
wood of that portion of the coffin immediately below his mouth
was stained a deep reddish-brown-black (*blood*).   Old Dr.
Donaldson, whom we were all very fond of, tried to explain
matters by saying that the jolting of the coffin on its way to the
cemetery had overturned the body, and that the blood stains on
the shirt and wood were the natural result of blood flowing (*i.e.*
oozing) out of the mouth of the corpse as it lay face downwards.
A nice theory, but scarcely a probable one, as all the jolting in
creation could not possibly turn a corpse over in an Indian coffin,
which is so built that there is scarcely two inches spare space over
any portion of the contained body, and unless the supposed corpse
regained consciousness and exerted *considerable* force, it could not
possibly turn round in its *narrow* casket.

"Mary Norah Best, aged seventeen years, an adopted daughter
of Mrs. C. A. Moore, *née* Chew, 'died' of cholera, and was entombed in the Chew's vault in the old French cemetery, at Calcutta.
The certifying surgeon was a man who would have benefited by
her death, and had twice (though ineffectually) attempted to put an
end to her adopted mother, who fled from India to England after
the second attempt on her life, but, unfortunately, left the girl
behind.   When Mary 'died' she was put into a *pine* coffin, the
lid of which was *nailed,* not screwed, down.   In 1881, ten years
or so later, the vault was unsealed to admit the body of Mrs.
Moore's brother, J. A. A. Chew.   On entering the vault, the undertaker's assistant and I found the lid of Mary's coffin on the floor,
while the position of the skeleton (half in, half out of the coffin,
and an ugly gash across the right parietal bone) *plainly* showed
that after being entombed Mary awoke from her trance, struggled
violently till she wrenched the lid off her coffin, when she either
fainted away with the strain of the effort in bursting open her

casket, and while falling forward over the edge of her coffin struck
her head against the masonry shelf, and died almost immediately;
or, worse still,—as surmised by some of her clothing which was
found hanging over the edge of the coffin, and the position of her
right hand, the fingers of which were bent and close to where
her throat would have been had the flesh not rotted away,—she
recovered consciousness, fought for life, forced her coffin open, and,
sitting up in the pitchy darkness of the vault, went mad with
fright, tore her clothes off, tried to throttle herself, and banged her
head against the masonry shelf until she fell forward senseless
and dead."

Dr. Chew says:—"Though a layman, still it would be
hard to find a more indefatigable sanitarian than my
late commanding officer, Lieutenant - Colonel R. C.
Sterndale, of the Presidency Volunteer Rifle Battalion,
and for many years vice-chairman of the municipality of
the suburbs of Calcutta. In order to prove his theory
that a great deal of danger existed in the rainy season
from subsoil water rising up into the graves, saturating
the bodies, and then poisoning the neighbouring tanks
and wells, he caused a trench, ten feet long, six deep,
and four wide, to be dug across an old Mahomedan
grave-yard. Soundings and measurements having been
taken of the subsoil water, he had a tarpaulin stretched
over the trench, and daily measured the 'fall' of the
water-level. He had a drawing made of the section of
that grave-yard in which the action of the nitre-laden
water seemed to mummify some of the bodies. Amongst
the rest was a somewhat mummified male corpse which,
instead of being on his back, was lying on his abdomen;
the left arm supported the chin, but had a piece of it
missing; the right hand clutched the left elbow, and the
general position of the body was as if, consciousness

having returned, the alleged corpse sat up, found the weight of the earth too heavy to work through, and then, dying of suffocation, fell forward in the position in which it was found and exposed."

Dr. Chew adds :—" I have heard and read of several other instances, but, as they have not come within my personal observation, I do not mention or refer to them."

CHAPTER V.

ALMOST every intelligent and observant person you converse with, if the subject is introduced, has either known or heard of narrow escapes of premature burial within his or her own circle of friends or acquaintances ; and it is no exaggeration to say that such cases are numbered by thousands. It is to be hoped that the number of timely discoveries vastly exceed those actually interred in a state of suspended animation ; but as no investigation of grave-yards or cemeteries (which effectually conceal their own tragedies) has ever taken place in England until the remains are reduced to dust, and rarely in other countries, one cannot be sure that this optimistic view is correct. The following cases of narrow escape appear to rest upon trustworthy evidence.

An apparent suspension of life, following a serious illness, is usually considered a satisfactory proof of the reality of the expected death ; but these conditions cannot always be relied upon. Cases are on record where the objects of such simulacra of death appear, if let alone, to gather the essence of renewed vitality, and return to consciousness. The *Undertakers' and Funeral Directors' Journal* of May, 1888, has a case in point.

" Mrs. Lockhart, of Birkhill, who died in 1825, used to relate to her grandchildren the following anecdote of her ancestor, Sir William Lindsay, of Covington, towards the close of the seven-

teenth century :—'Sir William was a humorist, and noted, moreover, for preserving the picturesque appendage of a beard at
a period when the fashion had long passed away. He had been
extremely ill, and life was at last supposed to be extinct, though,
as it afterwards turned out, he was merely in a "dead faint" or
trance. The female relatives were assembled for the "chesting"—
the act of putting a corpse into a coffin, with the entertainment
given on such melancholy occasions—in a lighted chamber in the
old tower of Covington, where the "bearded knight" lay stretched
upon his bier. But when the servants were about to enter to
assist at the ceremonies, Isabella Somerville, Sir William's great-
granddaughter, and Mrs. Lockhart's grandmother, then a child,
creeping close to her mother, whispered into her ear, "The beard
is wagging ! the beard is wagging !" Mrs. Somerville, upon this,
looked to the bier, and observing indications of life in the ancient
knight, made the company retire, and Sir William soon came out
of his faint. Hot bottles were applied and cordials administered,
and in the course of the evening he was able to converse with his
family. They explained that they had believed him to be actually
dead, and that arrangements had even been made for his funeral.
In answer to the question, "Have the folks been warned ?" (i.e.,
invited to the funeral) he was told that they had—that the funeral
day had been fixed, an ox slain, and other preparations made for
entertaining the company. Sir William then said, "All is as it
should be ; keep it a dead secret that I am in life, and let the
folks come." His wishes were complied with, and the company
assembled for the burial at the appointed time. After some delay,
occasioned by the non-arrival of the clergyman, as was supposed,
and which afforded an opportunity of discussing the merits of the
deceased, the door suddenly opened, when, to their surprise and
terror, in stepped the knight himself, pale in countenance and
dressed in black, leaning on the arm of the minister of the parish
of Covington. Having quieted their alarm and explained matters,
he called upon the clergyman to conduct an act of devotion, which
included thanksgiving for his recovery and escape from being
buried alive. This done, the dinner succeeded. A jolly evening,
after the manner of the time, was passed, Sir William himself
presiding over the carousals.'"

Dr. J. B. Vigné, in his "Memoire sur les Inhumations Précipitées," Paris, 1839, narrates the following:— "Mr. B., an inhabitant of Poitiers, fell suddenly into a state resembling death ; every means for bringing him back to life were used without interruption ; from continued dragging, his two little fingers were dislocated, and the soles of his feet were burnt; but, all these having produced no sensation in him, he was thought decidedly dead. As they were on the point of placing him in his coffin, some one recommended that he should be bled in both arms and feet at the same time, which was immediately done, and with such success that, to the astonishment of all, he recovered from his apparent state of death. When he had entirely recovered his senses, he declared that he had heard every word that had been said, and that his only fear was that he would be buried alive."

APPARENT DEATH IN PREGNANCY.

Hufeland (one of the greatest authorities on the subject in Germany), in his essay upon the uncertainty of the signs of death, tells of a case of the wife of Professor Camerer, of Tübingen, who was hysterical, and had a fright in the sixth month of her pregnancy, which brought on convulsions (eclampsia), which continued for four hours, when she seemed to die completely. Two celebrated physicians, besides three others of less note, regarded the case as ended in death, as all the recognised signs of death were present. However, attempts to revive her were at once resorted to, and were continued for five hours, when all the medical attendants, except one, gave the case up, and left. The physician who remained pulled off a blister-plaster that had been put on one of the feet, when the lady gave feeble signs of life by twitchings about the mouth. The doctor then renewed his efforts to revive her, by various stimulating means, and by burning, and by pricking the spine; but all in vain, for after her slight evidences of revival, she seemed to die unmistakably. She lay in a state of apparent death for six days, but there was a small space over the heart where a little warmth could be detected by the hand, and on this account the burial was put off. On the seventh day she opened her eyes, and slowly revived, but was completely unconscious of all that had happened. She then gave birth to a dead child, and soon thereafter recovered her health completely.

From the *Lancet*, November 27, 1858, p. 561.

"THE DEAD ALIVE.

" It seems to be always desirable to obtain a contemporary record of all unusual phenomena. It is so more especially where they are of a somewhat indefinite character, and scarcely susceptible of exaggeration. We know of none which are more so than the cases of 'trance.' These examples are both sufficiently unusual to deserve a passing record, and sufficiently mysterious in their character to call for a more careful investigation than it has hitherto been possible to accord to them. We transcribe the facts of a recent instance, as they are circumstantially detailed, and, no doubt, some of the surgeons of Coventry will be able to afford their testimony as to the degree of correspondence of this narrative with their observations.

" The girl, whose name is Amelia Hinks, is twelve or thirteen years of age, and resides with her parents in Bridge Street, Nuneaton. She had lately appeared to be sinking under the influence of some ill-explained disorder, and about three weeks since, as her friends imagined, she died. The body was removed to another room. It was rigid and icy cold. It was washed and laid out with all due funeral train. The limbs were decently placed, the eyelids closed and penny-pieces laid over them. The coffin was ordered. For more than forty-eight hours the supposed corpse lay beneath the winding-sheet, when it happened that her grandfather, coming from Leamington to assist in the last mournful ceremonies, went to see the corpse. The old man removed a penny-piece, and he thought that the corpse winked! There was a convulsive movement of the lid. This greatly disturbed his composure; for, though he had heard that she died with her eyes open, he was unprepared for this palpebral signal of her good understanding with death. A surgeon is said to have been summoned, who at first treated the matter as a delusion, but subsequently ascertained stethoscopically that there was still slight cardiac pulsation. The body was then removed to a warm room, and gradually the returning signs of animation became unequivocal. When speech was restored, the girl described many things which had taken place since her supposed death. She knew who had closed her eyes and placed the coppers thereon. She also heard the order given for her coffin, and could repeat the various remarks made over her as she lay in her death-clothes. She refused food, though in a state of extreme debility. She has since shown symptoms of mania, and is now said to have relapsed into a semi-cataleptic condition. The parents are 'creditable people,' and there is no apparent *ruse* in this unusually romantic history, which is causing considerable excitement in Nuneaton and its neighbourhood."

From the *Lancet*, December 18, 1858, p. 642.

"'THE DEAD ALIVE.'

"(To the Editor of the *Lancet*.)

"Sir,—An article, 'The Dead Alive,' in your impression of the 27th ultimo, demands of me a veritable statement of the case alluded to. The subject of the inquiry is still living, and for some time past has afforded me scope for observation.

"I have only been waiting for a termination of the case, either in convalescence or death, to enable me to give to the profession, through your valuable columns, a full and truthful history of this rare and curious case, replete with interest. The exaggerated statement which has gone the round of the press has produced such great curiosity in this immediate neighbourhood that I have been applied to by many parties, professional and non-professional, to be permitted to see the case, the parents of the patient having refused admittance to all strangers.

"The case having extended over a long period, and fearing a detailed account might occupy too much of your valuable space, I have condensed the matter as much as possible; but should the profession consider the case worthy of a more enlarged history, I will gladly, at some future period, meet their wishes, as far as my rough notes, aided by my memory, will supply it.

"In August, 1858, I was requested to visit Miss Amelia Hinks, aged twelve years and nine months, daughter of a harness-maker, and residing with her parents in Bridge Street, Nuneaton. She was supposed to be suffering from pulmonary consumption. . . . On October 18, about half-past three a.m., she apparently died. She is said to have groaned heavily, waved her hands (which was a promised sign for her mother to know that the hour of her departure was come), turned her head a little to the light, dropped

6

her jaw, and *died.* In about half an hour after her supposed departure she was washed, and attired in clean linen, the jaw was tied by a white handkerchief, penny-pieces laid over her eyes, her hands, semi-clenched, placed by her side, and her feet tied together by a piece of tape. She was then carried into another room, laid on a sofa, and covered over with a sheet. She appeared stiff and cold, two large books were placed on her feet, and I have no doubt she was considered to be a sweet corpse.

"About nine a.m., the grandfather of the supposed dead went into the death-chamber to give a last kiss to his grandchild, when he fancied he saw a convulsive movement of the eyelid, he having raised one of the coins. He communicated this fact to the parents and mourning friends, but they ridiculed the old man's statement, and said the movement of the eyelids was owing to the nerves working after death. Their theory, however, did not satisfy the experienced man of eighty years, and he could not reconcile himself to her death. As soon as I reached home, after having been out in the country all night, I was requested to see the child, to satisfy the old man that she was really dead. About half-past ten a.m. I called ; and immediately on my entrance into the chamber I perceived a tremulous condition of the eyelids, such as we frequently see in hysterical patients. The penny-pieces had been removed by the grandfather. I placed a stethoscope over the region of the heart, and found that organ performing its functions perfectly and with tolerable force. I then felt for a radial pulse, which was easily detected, beating feebly, about seventy-five per minute. The legs and arms were stiff and cold, and the capillary circulation was so congested as at first sight to resemble incipient decomposition. I carefully watched the chest, which heaved quietly but almost imperceptibly ; and immediately unbandaged the maiden, and informed her mourning parents that she was not dead. Imagine their consternation ! The passing-bell had rung, the shutters were closed, the undertaker was on his way to measure her for her coffin, and other necessary preparations were being made for her interment. [The writer then proceeds to give interesting details as to the treatment of the case, and the means taken to promote recovery.]

"RICHARD BIRD MASON, M.R.C.S., L.S.A.
"Bridge Street, Nuneaton, December 14, 1858."

From the *Lancet*, March 5, 1859, p. 254.

"TRANCE.

"Another case of trance is reported, in addition to those which we have lately recorded. A widow named Aufray, about sixty years of age, of St. Agnan de Cenuières (Eure), long seriously ill, became suddenly worse, grew cold and motionless, and, as it was thought, dead. She was laid out, the coffin ordered, and the church bell tolled. She recovered consciousness just before the funeral was to take place."

THE QUESTION OF PREMATURE BURIAL BEFORE THE FRENCH SENATE.

The *Medical Times*, London, 1866, vol. i., p. 258, under the heading "Buried Alive" remarks as follows:—"The abundance of other topics hinders us at present from saying more than a few words on the conditions under which there may be real danger of burial before life is quite extinct. Now, we will only reproduce the cases reported by Cardinal Archbishop Donnet, in the French Senate, in a discussion on a petition that the time between death and burial should be lengthened. We will add one instance, which we have heard on the best authority:—About thirty years ago, a young woman of eighteen, daughter of Madame Laligand, living in the Rue des Tonnelliers, at Beaune, in Burgundy, was supposed to have died. The ordinary measures were taken for interment. The body was put in a coffin, and taken to the church ; the funeral service was said, and the *cortége* set out for the cemetery ; but on the road between the church and the cemetery the supposed dead recovered power of motion and speech, was removed from the coffin, put to bed, recovered, married, and lived eighteen years afterwards. She said

she retained her consciousness during the whole of her
supposed death, and had counted the nails that were
driven into her coffin.   Statements such as these, and
such as those made by the Archbishop, will surely be
subjected to the ordeal of a French scientific commission, and we may suspend our judgment for the present.
To return to his Eminence.   He said he had the very
best reasons for believing that the victims of hasty
interments were more numerous than people supposed.
He considered the rules and regulations prescribed by
the law very judicious ; but, unfortunately, they were,
particularly in the country, not always executed as they
should be, nor was sufficient importance attached to
them.   In the village he was stationed in as an assistant-
curate in the first period of his sacerdotal life, he saved
two persons from being buried alive.   The first an
aged man, who lived twelve hours after the hour prescribed for his interment by the municipal officer ; the
second was a man who was quite restored to life.   In
both cases a trance more prolonged than usual was taken
for actual death.   The other instances, says the *Times'*
correspondent, I give in the words of the Archbishop :—

"'The next case that occurred to me was at Bordeaux.   A
young lady, who bore one of the most distinguished names in
the Department, had passed through what was supposed the last
agony, and, as apparently all was over, the father and mother
were torn away from the heartrending spectacle.   As God willed
it, I happened to pass the door of the house at the moment,
when it occurred to me to call and inquire how the young lady
was going on.   When I entered the room, the nurse, finding the
body breathless, was in the act of covering the face, and, indeed,
there was every appearance that life had departed.   Somehow or
other, it did not seem to me so certain as to the bystanders.   I
resolved to try.   I raised my voice, called loudly upon the young

lady not to give up all hope—that I was come to cure her, and that I was about to pray by her side. "You do not see me," I said, "but you hear what I am saying." My presentiments were not unfounded. The word of hope I uttered reached her ear and effected a marvellous change, or, rather, called back the life that was departing. The young girl survived ; she is now a wife, and mother of children, and this day is the happiness of two most respectable families.'

"The Archbishop mentioned another instance of a similar revival in a town in Hungary during the cholera of 1831, which he heard that day from one of his colleagues of the Senate, as they were mounting the staircase. But the last related is so interesting, and made such a sensation, that it deserves to be repeated in his own words :—

"'In the summer of 1826, on a close summer day, in a church which was exceedingly crowded, a young priest, who was in the act of preaching, was suddenly seized with giddiness in the pulpit. The words he was uttering became indistinct ; he soon lost the power of speech, and sank down on the floor. He was taken out of the church and carried home. All was thought to be over. Some hours after, the funeral bell was tolled, and the usual preparations made for the interment. His eyesight was gone : but if he could see nothing, like the young lady I have alluded to he could hear, and I need not say that what reached his ears was not calculated to reassure him. The doctor came, examined him, and pronounced him dead ; and after the usual inquiries as to his age and the place of his birth, etc., gave permission for his interment next morning. The venerable bishop, in whose cathedral the young priest was preaching when he was seized with the fit, came to his bedside to recite the "De Profundis." The body was measured for the coffin. Night came on, and you will easily feel how inexpressible was the anguish of the living being in such a situation. At last, amid the voices murmuring around him, he distinguished that of one whom he had known from infancy. That voice produced a marvellous effect and superhuman effort. Of what followed I need say no more than that the seemingly dead man stood next day in the same pulpit. That young priest, gentlemen, is the same man who is now speaking before you, and who, more than forty years after that event, implores those in authority, not merely to watch vigilantly over the careful execution of the legal prescriptions with regard to interments, but to enact fresh ones in order to prevent the recurrence of irreparable misfortunes.'"

To this report of the *Medical Times* it may be added that the petition of M. de Carnot furnished statistics showing the frequency of these terrible disasters, and suggested various preventive measures, including the establishment of mortuaries, a longer interval between death and burial, and the application of scientific methods of restoration where decomposition is not manifest. The reality of the terrible dangers, as pointed out by Cardinal Donnet, was confirmed by Senators Tourangin and Viscount de Baral, in the recital of other cases of premature interment.

When the subject was revived in the Senate on January 29, 1869,—on which occasion five petitions were presented, urging important reforms, and detailing other cases of premature interment,—Cardinal Donnet again took part in the debate, and urged that no burial should be permitted without the signature of a doctor or officer of health, as well as the written authorisation of the Mayor, so that the fact of death might always be verified. The Cardinal then furnished particulars of another recent case of premature interment in l'Est, and recalled the fact that one of their honourable colleagues of the Senate, M. le Comte de la Rue, had had a narrow escape from live sepulture.

The several petitions were forwarded to the Minister of the Interior, but nothing was done to remedy the evil. From the *Lancet*, June 2, 1866, p. 611.

"ON SUSPENDED ANIMATION.

" In the course of the address delivered by Dr. Brewer to the Guardians of St. George's at St. James's Hall, he adverted to the 'laying-out' case at St. Pancras. . . . Dr. Brewer . . . dwelt upon the question of suspended animation in a passage which really deserves to be quoted. . . .

"' I have been more than once under a condition of apparently suspended respiration, and with circumstances less comfortable than those related of this babe ; and yet, active as is my brain, and sensitive as is my body, I remember as well as though it were but yesterday that, on being restored to consciousness, no feeling of discomfort of any kind attended my experience on either occasion. It is under the truth to say I have known a score of cases of those who have been supposed dead being reanimated. It is not many months ago a friend of mine, a rector of a suburban parish, was pronounced by his medical attendant to be dead. His bed was arranged, and the room left in its silence. His daughter had re-entered and sat at the foot, and the solemn toll of his own church bell was vibrating through the chamber, when a hand drew aside the closed curtain, and a voice came from the occupant of the bed—"Elizabeth, my dear, what is that bell tolling for?" The daughter's response was, perhaps, an unfortunate one : " *For you, papa.*" Schwartz, the first eminent Indian missionary, was roused from his supposed death by hearing his favourite hymn sung over him previous to the last rites being performed, and his resuscitation made known by his joining in the verse.'"

Dr. B. W. Richardson quotes a case in the *Lancet*, 1888, vol. ii., p. 1179, of a man who, in 1869, was rendered cataleptic by a lightning - stroke, and who narrowly escaped living burial.

Dr. Moore Russell Fletcher in his work on "Suspended Animation," p. 26, says:—

"In June, 1869, a girl in Cleveland, Ohio, was taken ill, and after a short sickness died, and was laid out for burial; but as her mother insisted that she was not dead, efforts were made for some time to restore her to life, but in vain. Her mother, however, refused to let her be buried; and on the fifth day after that set for the funeral the slamming of a door aroused her, so that she recovered. She stated that, during most of the eight days which she lay there, she was conscious and heard what was said, although wholly unable to make the least motion."

Dr. M. S. Tanner in a letter to the *New York Times*, January 18, 1880, mentions two cases where persons awakened from trance at the moment of sepulture described in turn what their feelings had been. Said one :—

"Have you ever felt the paralysing influence of a horrible nightmare? If you have had such experience, then you are prepared to conceive of the mental agonies I endured when I realised that my friends believed me dead, and were making preparations for my burial. The hours and days of mental struggle spent in the vain endeavour to break loose from the vice-like grasp of this worse than horrible nightmare was a hell of torment such as no tongue can describe or pen portray."

The other instance mentioned by Dr. Tanner is that of Dr. Johnson of St. Charles, Illinois, who in the hearing of Dr. Tanner, and in the presence of a large audience in Harrison's Hall, Minneapolis, stated that when a young man he was prostrated with a fever. He swooned away, apparently dead. His attending physician said he was dead. His father was faithless and unbelieving, and refused to bury him. He lay in this condition, apparently dead, fourteen days. The attending physician brought other physicians to examine the apparently lifeless form, and all stated unqualifiedly, " He is dead." Some fourteen physicians, among them many eminent professors, examined the body, and there was no ambiguity in the expression of their conclusion that the boy was dead. But the father still turned a deaf ear to all entreaties to prepare the body for the grave. Public feeling was at last aroused. The health officer and other city officers, acting in their official capacity, and by the advice of physicians, peremptorily demanded that the body be interred without delay. On the fourteenth day the father yielded under protest ; preparations were made for the funeral, when the emotions of the still living subject, who was conscious of all transpiring around him, were so intense as to be the means of his deliverance. He awoke from his trance.

From the *Lancet*, June 7, 1884, p. 1058.

"IMPORTANT SUGGESTION FROM AN M.D.

" (To the Editor of the *Lancet*.)

" Sir,—Without venturing to express an opinion on the case mentioned by the Rev. D. Williams[1] in the *Lancet* of the 24th inst., I would beg to say that I have no doubt in my own mind but that people are sometimes 'buried alive.' An instance has come to my knowledge where this catastrophe was only avoided by a mere accident. A lady, about forty-five years of age, the wife of a clergyman in a northern county, was taken ill, and after some time, as was supposed, died.

---

[1] The case referred to, being attended with considerable doubt, is omitted.

The funeral was delayed, and so was the closing of the coffin, in consequence of the absence of a son of the lady from home. When the boy arrived, the kissing, wailing, and commotion roused the supposed dead woman, and brought her to consciousness in her coffin. This lady would most probably have been buried alive were it not that the obsequies were delayed on account of the circumstance mentioned.

"Now, may not cases more or less similar to this sometimes occur, with the catastrophe of 'buried alive' added to them? But no such case could happen if it were made compulsory that the interment of a body should not be allowed to take place until after decomposition had set in, as attested by a medical man.

"I am, Sir, yours truly,

"WM. O'NEILL, M.D.

"Lincoln, May 26, 1884."

It is not always safe to conclude that persons enfeebled by age, or exhausted by long and severe illness, and pronounced dead by the attendant doctor, are really so. The *Undertakers' Journal*, August 23, 1886, has the following :—

"It appears that George O. Daniels, of Clinton, Kentucky, had been ill for several months, and at length, to all appearance, died. The body was put in a coffin, where it remained for twenty hours, awaiting the arrival of relatives to attend the funeral. At midnight the watchers who surrounded the coffin were startled by a deep groan emanating from it, and all but one, a German of the name of Wabbeking, rushed from the room. Wabbeking remained, and as the groans continued he raised the coffin-lid and saw that Daniels was alive. Seizing the body he placed it upright. A few spasmodic gasps, a shudder, and the corpse spoke. The relatives returned to find the man sitting in a chair, and conversing with reasonable strength. Mr. Daniels claims to have been perfectly conscious of everything which passed around him, but says he was unable to move a muscle. He heard the sobs of his relatives when he was pronounced dead by the doctors, and noticed the preparations for the funeral. He is about eighty years of age."

The same journal for July 23, 1888, reports the following under the head of

"RETURNED TO LIFE TWICE.

" The following details are given by the Cincinnati correspondent of the *New York Herald* from Memphis, Tennessee :—Mrs. Dicie Webb keeps a grocery store on Beale Street, and is well known to hundreds. Two years ago John Webb, a son of Mrs. Webb, married Sarah Kelly, a pretty girl, to whom the mother-in-law became greatly attached. Before one year of their married life had passed, Mrs. Webb, jun., was stricken with consumption, and on several occasions came near dying. About a month ago the young woman became very anxious to visit her parents in Henderson County, and she was taken there. At first she appeared much improved, and hopes were felt that her life might be preserved through the summer, but two weeks ago last Tuesday a telegram announced her death, and the husband hurried to her parents' home. Three days later he returned with the corpse. The mother-in-law pleaded so hard for a sight of the dead woman, that finally, despite the belief that the body was badly decomposed, it was decided to open the coffin. While looking at the placid face Mrs. Webb was terrified at beholding the eyelids of the dead woman slowly opening. The eyes did not have the stony stare of death, nor the intelligent gleam of life. Mrs. Webb was unable to utter a sound. She could not move, but stood gazing at the gruesome sight. Her horror was increased when the supposed corpse slowly sat upright and, in an almost inaudible voice, said, ' Oh, where am I ?' At this the weeping woman screamed. Friends who rushed into the room were almost paralysed at the sight, and fled shrieking. But one bolder than the others returned and spoke to the woman, who asked to be laid on the bed. Hastily she was taken from the coffin and cared for. In the course of the day the resurrected woman fully regained her mental powers. The day following she related a wonderful story. She said she was cognisant of all that occurred, and did not lose consciousness until she was put aboard the train for Memphis. Soon after being placed in her mother-in-law's house she came to her senses and knew all that was passing. While her mother-in-law was looking at her she made a supreme effort to speak. Mrs. Webb lived a number of days, when she again apparently died. The doctors pronounced her dead. and she was once more placed in the coffin. While the mother-in-law was taking her final farewell she heard a voice whisper, ‘Mother. don't cry.’ Looking into the girl's face. she saw the same look that she had noticed before. She called for help, and several women responded. Some one cried, ‘Shake her ; she's not dead.’ In the excitement of the moment, the women, it is thought, shook the life out of the poor consumptive, and last Saturday she was buried. The family and friends have endeavoured to keep the matter quiet.”

The *Daily Telegraph*, January 26, 1889, reports :—

“A NARROW ESCAPE.

“A Rochester correspondent telegraphs that a woman named Girvin, living at Burham, near Rochester, has just had a narrow escape of being buried alive. She fell into a kind of trance, which was mistaken for death. The coffin was ordered, and the usual preparations made for a funeral. But while a number of the relatives were gathered at the bedside bewailing their bereavement, the supposed corpse startled them by suddenly rising up in bed and asking what was the matter. The woman is making good progress towards convalescence.”

And on July 6, 1889, the same journal says :—

“Our St. Leonards correspondent telegraphs :—About a week ago the wife of a well-known tradesman in St. Leonards fell ill, and on Monday night last the doctor gave his opinion that she could not live through the next day. On Tuesday morning at ten o'clock the doctor pronounced his patient dead, the nurse who was in attendance confirming the opinion. The intimation of death naturally created great distress among the friends of the woman, who was laid out in grave-clothes, washed, and prepared for burial, and, being a Roman Catholic, a crucifix was placed in her hand as she lay on her bier. When it was announced that the woman was dying, a priest was sent for ; but he could not attend, as he was out of the town at the time.

"About a quarter to ten on Tuesday night the nurse entered the room without a light for the purpose of getting something which she knew where to find. Whilst in the darkened chamber she was startled to hear a slight cry proceeding from the bed where the body lay, and she rushed from the room in a terrible fright. The widower, hearing the scream of fright, rushed into the chamber with a light, and was astounded to find that his wife had raised herself up in the bed on her elbow. She faintly uttered the words, 'Where am I?' and again relapsed into a heavy sleep. The opportunity was seized of changing the shroud for proper habiliments, and in about an hour and a half she woke again perfectly conscious. Next morning she was told of what had occurred, but was quite ignorant of everything that had passed, thinking she had only had a long sleep. She is now doing well, and it is hoped she will soon be restored to health and strength. The doctor describes the case as the most remarkable he has ever met with in his experience."

Dr. Frederick A. Floyer, of Mortimer, Berks, published the following case in the *Tocsin*, November 1, 1889, vol. i., p. 84, under the head of " Premature Burial":—

" A narrow escape of this was recently communicated direct to the writer, and as it has some extremely important bearings on the value of what are usually considered to be evidences of death, we give it as told by the survivor, who is still alive in the form of a cheery and intelligent old lady in the fullest possession of her faculties and memory.

" Herself the wife of a medical officer attached to the —th Regiment, she was stationed at —— Island, where at the age of twenty-eight she was safely confined. Shortly after this she was walking out with an attendant when she was taken suddenly ill with a painful spasm of the heart—what appears to have been an attack of angina pectoris—and was conveyed in-doors and propped up with pillows, suffering great pain, and although medical attendance was summoned, nothing was of avail, and she died—at least in the opinion of those around her, who paid the proper attention to what they regarded as a corpse. It was the custom there to bury at sundown any one who died during the day. We understand that in warm countries it is difficult to close the eyelids properly, and so this lady, lying motionless and rigid, contemplated with perfectly clear perception, but with an utter indifference, the bringing in of the coffin and the necessary preparations for her interment ; she remembers her children coming to take a last look at her, and then being taken down stairs.

" She would never have lived to tell the story but for an accident, which happened in this way. Her nurse, who was much attached to her, was stroking her face and the muscles of her jaw, and presently declared she heard a sound of breathing. Medical assistance was summoned, and the mirror test applied, but the surface was undimmed. Then, to make sure, they opened a vein in each arm, but no blood flowed. No limb responded to stimulus, and they declared that the nurse was mistaken, and that the body was dead beyond doubt.

"But the nurse persisted in her belief and in her attentions, and did succeed in establishing a sign of life. Then mustard applications to her feet and to the back of her neck, and burnt feathers applied to her nostrils, which she remembered burning her nose, completed her return to consciousness."

From the *Pall Mall Gazette*, May 11, 1891.

"NARROW ESCAPE FROM BEING BURIED ALIVE.

"A Penn Station telegram to Dalziel says :—A singular case of simulation of death from fright occurred here on Saturday. Mrs. Sarseville, the wife of a farmer in this county, was in the cow-house attending to the dairy work when she saw a nest of squirming snakes through a hole in the plank floor. She fell to the ground apparently lifeless with fright. Help was summoned, and she was carried into the house. Before the physician arrived Mrs. Sarseville had begun to turn black, and he pronounced her dead, giving a certificate, in which he assigned apoplexy as the cause. During the night Mrs. Sarseville's daughter sat beside the coffin of her mother, lamenting her death. Just before daybreak she was startled to see the body move. She was more shocked when her mother opened her eyes and sat bolt upright in her coffin. The supposed corpse was no less startled than the girl to find herself dressed in grave-clothes and lying in a coffin. Help was summoned, and the lady helped out of her narrow bed and into her ordinary clothes. She took breakfast with the family yesterday morning, and seemed none the worse for her ghastly experience."

From the *British Medical Journal*, March 12, 1892, p. 577.

"A NARROW ESCAPE FROM PREMATURE BURIAL.

" The *Temps* publishes a case of premature burial prevented by the daughter of the supposed dead man, who, on kissing her father, perceived that his body was not cold. The funeral *cortége* was on the point of starting. Suitable measures restored the man to consciousness, and he opened his eyes and uttered one or two words. His condition is serious, but he is alive. This incident occurred at Vagueray, near Lyons."

From the *Echo*, London, May 13, 1893.

"ALMOST BURIED WHILE ALIVE.

"Limoges, May 13.

"A woman has just had a narrow escape of being buried alive here. She was subject to epileptic fits, and during one of these a few days ago was pronounced to be dead. The arrangements for interment were made in due course, and as the coffin was being borne into the church some of the mourners said they heard a knocking inside. The party listened. and distinct taps were heard. No time was lost in wrenching off the lid of the coffin. It was then found that the woman was alive and conscious, although terribly frightened at the awful ordeal through which she had passed. A doctor was quickly in attendance, and under his direction the supposed corpse was removed from the coffin and placed on a litter for conveyance home again."

The *Undertakers' Journal*, July 22, 1893, says :—

"Charles Walker was supposed to have died suddenly at St. Louis a few days ago, and a burial certificate was obtained in due course from the coroner's office. The body was lying in the coffin, and the relatives took a farewell look at the features, and withdrew as the undertaker's assistants advanced to screw down the lid. One of the undertaker's men noticed, however, that the position of the body in the coffin seemed to have undergone some slight change, and called attention to the fact. Suddenly, without any warning, the 'corpse' sat up in the coffin and gazed round the room. A physician was summoned, restoratives were applied, and in half an hour the supposed corpse was in a warm bed, sipping weak brandy and water, taking a lively interest in the surroundings. Heart-failure had produced a species of syncope resembling death that deceived even experts."

From the *Undertakers' Journal*, August 22, 1893.

"SNATCHED FROM DEATH AT THE GRAVESIDE.

"A marvellous case of suspended animation is described from the British colony of Lagos, where an old woman named Oseni came to life when she was at the cemetery about to be buried. The mourners had assembled at the cemetery, and, in accordance with the Mahomedan rule, the body was lifted from the coffin to be buried, when several distinct coughs were given by the supposed corpse. She was at once released from the clothes which bound her, and the old woman, to the surprise and amazement of those present, sat upright and opened her eyes. Some gruel was then procured, of which she partook with evident relish."

From the *Daily Telegraph*, London, December 12, 1893.

"A LADY NEARLY BURIED ALIVE.

"Berlin, December 11.

"From Militsch, in Silesia, an extraordinary case of trance is reported. It seems that, owing to the grave not being in readiness, some delay occurred in the burial of a lady, the wife of a major in the army, who to all appearance had died. On the fourth day after the lady's supposed death the maid was placing fresh flowers round the coffin, when she was much startled at seeing the body move, and finally assume an erect position. The lady had evidently been in a state of coma during the past four days, and narrowly escaped being buried alive."

The *Banner of Light*, Boston, July 28, 1894, quotes the following case of apparent sudden death from the *Boston Post*:—

"COFFINED ALIVE!

"Sprakers, a village not far from Rondout, N.Y., was treated to a sensation Tuesday, July 10, by the supposed resurrection from the dead of Miss Eleanor Markham, a young woman of respectability, who to all appearance had died on Sunday, July 8.

"Miss Markham about a fortnight ago complained of heart trouble, and was treated by Dr. Howard. She grew weaker gradually, and on Sunday morning apparently breathed her last, to the great grief of her relatives, by whom she was much beloved. The doctor pronounced her dead, and furnished the usual burial certificate.

7

"Undertaker Jones took charge of the funeral arrangements. On account of the warm weather it was decided that the interment should take place Tuesday, and in the morning Miss Markham was put in the coffin.

"After her relatives had taken the last look on what they supposed was their beloved dead, the lid of the coffin was fastened on, and the undertaker and his assistant took it to the hearse waiting outside. As they approached the hearse a noise was heard, and the coffin was put down and opened in short order. Behold! there was poor Eleanor Markham lying on her back, her face white and contorted, and her eyes distended.

"'My God!' she cried, in broken accents. 'Where am I? You are burying me alive.' 'Hush! child,' said Dr. Howard, who happened to be present. 'You are all right. It is a mistake easily rectified.'

"The girl was then taken into the house and placed on the bed, when she fainted. While the doctor was administering stimulating restoratives the trappings of woe were removed, and the hearse drove away with more cheerful rapidity than a hearse was ever driven before.

.   .   .   .   .   .

"'I was conscious all the time you were making preparations to bury me,' she said, 'and the horror of my situation is altogether beyond description. I could hear everything that was going on, even a whisper outside the door, and although I exerted all my will-power, and made a supreme physical effort to cry out, I was powerless. . . . At first I fancied the bearers would not hear me, but when I felt one end of the coffin falling suddenly, I knew that I had been heard.'

"Miss Markham is on a fair way to recovery, and what is strange is that the flutterings of the heart that brought on her illness are gone."

From the *Echo*, January 18, 1895.

"MISTAKEN FOR DEAD—A WOMAN'S AWFUL EXPERIENCE.

"An extraordinary affair is reported from Heap Bridge, Heywood. Yesterday a woman was supposed to have died, and she was washed, laid out, and measured for her coffin, a piece of linen being placed over her mouth. Eight hours later, however, as two women were engaged in the room. the supposed corpse blew the linen away, and raised herself up in bed. The two women were terribly frightened, and in their hasty retreat both tumbled downstairs. and are now suffering from slight injuries, as well as shock. Some time elapsed before any one else could be induced to enter the house, but eventually several persons went in together, and found the woman still sitting up in bed. She was exceedingly weak. Later, however, she succumbed, and the doctor expressed the opinion that her death was accelerated by shock. During the night the woman conversed with her son, who had carried her upstairs for dead, and told him of the awful sensation she felt whilst unable to speak during the washing and laying out of her body."

The following letter appeared in the London *Daily Chronicle* of September 24, 1895 :—

" BURIED ALIVE.

·" Sir,—To your interesting correspondence on ' Buried Alive,' I would add the following, which I had directly from the mouth of one who but for the faithfulness of her husband would probably have been added to the number. I knew her quite well. She was the daughter of a physician in my native town, and her husband was a professor of music, and I will tell the incident as nearly as I can remember in her own words. She said :—' I had in my early married life a dread of there being any mistake made about my death. and begged my husband that, should he survive me, he would watch my body himself, which he promised he would do. Some time after this, I was overtaken by a most terrible attack of fever, succeeded by entire exhaustion, and I, as my attendants believed, died, and was accordingly laid out for burial. My good husband was true to his promise, and he, with my sister, watched the corpse, and in the night they perceived some indication of returning life, and of course means were used for restoration.'

" I cannot be quite sure how many years she lived after, but she had brought up at the time I speak of a family of four sons and one daughter, and she lived to a good old age.—Yours truly,

" CASSANDRA M——.

" September 18."

Speaking on the subject of premature burial the other day, a well-known London publisher told the author that he personally knew a lady, the daughter of a British Consul, who had been taken for dead on two separate occasions. On the first occasion the lady had been placed in her coffin, and the lid screwed down ready for interment. A friend who had known the supposed deceased called to condole with the family, and said :—" I should like to have a last look at dear L—— if you will only permit me." The lid was accordingly removed, and the visitor detected, as it seemed to her, signs of life in her friend ; she was taken out of her coffin, put in a warm bath, and recovered. Some years later the same lady fell into a cataleptic state after a fever, and was taken for dead. Preparations had been made for the funeral in both instances, but delayed beyond the usual time for interment. She returned to consciousness, and is now living.

Dr. Moore Russell Fletcher in " Suspended Animation and the Danger of Burying Alive," p. 62, writes :—

" ' Seven hours in a coffin added ten years to my life,' was the remark of Martin Strong, of Twelfth Street, Philadelphia, some time after quitting the coffin in which his family had placed him for burial, after Dr. Cummings had given a certificate of his death. Frank Stoop, of Clarinda, Iowa, was laid out for burial not long since, a physician having certified to his death ; but fortunately he awoke from his state of coma in time to save his life."

AN ARMY SURGEON'S PERSONAL EXPERIENCE.

Dr. R. G. S. Chew, of Calcutta, writing to the author, says :—" In 1873 I was a student in the Bishop's High School, Poonah (Bombay Presidency), where I used to be generally at the head of my class, and when competing for the Science Prizes I was fully determined to take the first prize or none. The Reverend —— Watson, Rector of St. Mary's Church and Chaplain to our school, knew my disposition, and cautioned me against being too sanguine, lest disappointment might tell very keenly. The disappointment came, and with it much nervous excitability. Shortly after this (Christmas, 1873) my favourite sister was seized with convulsions that carried her off. From the moment of her decease to nearly a month after her interment I entirely lost the power of speech. On the day of the funeral I was parched with thirst, but could not drink, as the water seemed to choke me. My eyes were burning and my head felt like bursting, but I could neither sob nor cry. I felt quite dazed, and followed the procession to the cemetery, where I stood motionless by the open grave ; but as soon as they lowered the little coffin into its resting-place I threw myself headlong into the grave and fainted away. Some one pulled me out and carried me home, where I lay in a sort of stupor for nine days, during which Dr. Donaldson attended me most patiently, and I regained consciousness, but was too weak to even sit up in bed. On the 16th January, 1874, I felt a peculiar sensation as of something filling up my throat—no swelling, no pain nor anything that pointed to throat affection —and this getting worse and worse, in spite of everything,

I *died*, as was supposed, on the 18th of January, 1874. and was laid out for burial, as the most careful examination failed to show the slightest traces of life. I had been in this state for twenty hours, and in another three hours would have been closed up for ever, when my eldest sister, who was leaning over the head of my coffin crying over me, declared she saw my lips move. The friends who had come to take their last look at me tried to persuade her it was only fancy, but,as she persisted, Dr. Donaldson was sent for to convince her that I was really dead. For some unexplained reason he had me taken out of the coffin and examined very carefully from head to foot. Noticing a peculiar, soft fluctuating swelling at the base of my neck, just where the clavicles meet the sternum, he went to his brougham, came back with his case of instruments, and, before any one could stop him or ask what he was going to do, laid open the tumour and plunged in a tracheotomy tube, when a quantity of pus escaped, and, releasing the pressure on the carotids and thyroid, was followed by a rush of blood and some movement on my part that startled the doctor. Restoratives were used, and I was slowly nursed back to life ; but the tracheotomy tube (I *still* carry the scar) was not finally removed till September, 1875."

"APPARENT DEATH FROM A FALL.

(*Communicated to the author by Dr. Chew.*)

"A sowar—*i.e.*, native trooper—of the 7th regiment of cavalry, in 1878, carrying despatches at Nowshera. was thrown from his horse, and, falling with his head against a sharp stone in the road, rolled on to his back, in which position he was found some six or

seven hours after, and conveyed to the morgue of
the European Depôt Hospital pending removal to
the 'lines' of his own corps. There was very little
hæmorrhage, and the stone was still wedged in
between the temporo-parietal suture. Cardiac sounds
and respiratory murmurs could not be detected. The
limbs were perfectly rigid, and there was a good deal of
cadaveric ecchymosis to be distinctly seen. Nothing
would have convinced any one that the sowar was still
alive, and Surgeons-Major Hunter, Gibson, and Briggs,
Apothecary S. Pollock, Assistant-Surgeon J. Lewis and
myself *verily* believed he was stone-dead. As 'cause of
death' is what the army is exceedingly particular about,
Surgeon-Major Hunter removed the impacted stone
and lifted out portions of the fractured bone (prior to
holding a proper *post-mortem*), when to the surprise of
all of us 'the corpse' deliberately closed its eyes (which
were staring open when the body was first brought in),
and there was a slight serous hæmorrhage. On noticing
this, the sowar's head was trephined,—no chloroform or
other anæsthetic being used,—some more fragments of
bone and a large blood-clot that pressed on the brain
were removed, and, as the sowar repeatedly flinched
under this operation, a stimulant was poured down his
throat, and he was removed to his regimental hospital,
from which he was discharged 'well' some six months
and a half later. After this he did good service in the
Afghan and Egyptian campaigns."

"APPARENT DEATH FROM CHOLERA.

"The cases of collapse and apparent death during
epidemics of cholera are very numerous, as will be seen

by reference to medical literature. We have now before us particulars of cases from the *Calcutta Journal of Medicine* for 1869, vol. ii., p. 383, where Dr. Charles Londe, of Paris, observes that patients pronounced dead of cholera have been repeatedly seen to move. See also, for Italy, *Lancet*, 1884, vol. ii., p. 655.

" A correspondent, signing himself T.E.N., in *To-Day*, October 12, 1895, says :—' When acting as special correspondent to the *Evening Herald* in Hamburg during the cholera plague, I met a gentleman who had been passed for dead and placed in the mortuary to await burial. When the porters entered some hours later to remove the hundred or so bodies, they found this gentleman sitting up in great pain, and very much frightened. He was placed in a ward and recovered. About the same time a little girl came to life actually at the graveside. She had been brought in one of several four-horse vans that conveyed bodies for interment in the Ohlsdorff graveyard. Fortunately for her, she had not been placed in a coffin, the exigencies of the time rendering it impossible to provide caskets for the dead. When the disease began to die out, the people found time to ask—" Can it be possible that life remains in any of the bodies buried?" That the doctors in the latter days cut the ulnar arteries of all subjects before passing them for dead is full of significance.'"

The three following cases were communicated to the author, during his sojourn in Calcutta, by Dr. Chew, in the early part of this year (1896) :—

" In March, 1877, Assistant-Surgeons H. A. Borthwick, S. Blake, H. B. Rogers, and myself received orders to proceed from Rawal Pindi by bullock-train to Peshawur to join the various regiments we were to be posted to for duty. We had just passed a place called Rati when Borthwick showed strong symptoms of cholera, from which he suffered all that night. The nearest hospital was twenty-five miles behind us, and though we had neither medicines nor sick-room comforts with us, we had no alternative but to journey onwards, because the train-drivers (Indians) refused to turn back, and if we did return to Rawal Pindi we would have been court-martialled for disobeying lawful commands and coming back without orders to do so. Travelling by bullock-train is very slow work, and far from a comfortable mode of transit; however, we were obliged to make the best of it, and early next morning Borthwick was cold, stiff, and seemingly dead. Here was a fine state of affairs —the nearest cantonment, which we had no expectation of reaching (*i.e.*, Nowshera) before nine p.m., was thirty-six miles off, and by the time we arrived at it, it would have been too late to approach the authorities, while Peshawur, our destination, was another twenty-nine miles further off. Dispose of the body we dared not, and we had no choice but to continue our route. All that day there was not a movement or other sign to show that life was not extinct, and affairs seemed no better by five p.m. next day, when we reached Peshawur. The apparent corpse was lifted out of the bullock-train and carried into the hospital dispensary (where a strong fire was blazing) preparatory to papers being signed and arrangements made for its final disposal. Whether it was the heat of the fire before which he was placed, or whether the vibriones had produced an antitoxin, I am not prepared to argue; but *we do know* that Borthwick recovered consciousness while lying on the bed in that dispensary, and that he whom we mourned as dead returned to life. He served in the same military stations with me in the North-West Frontier till 1880, when he accompanied me to the Calcutta Medical College, where we parted company in February, 1882, I bound for Egypt and he for frontier duty. At first we corresponded regularly, but since 1885 we lost touch of each other."

"REVIVAL IN A MORTUARY IN INDIA.

"Sergeant J. Clements Twining, of H.M.'s 109th regiment of British infantry, located at Dinapoor in 1876, was brought in an unconscious state to the hospital, supposed to be suffering from *coup de soleil*. Everything that could be done was ineffectually tried to rouse him from coma, and he was removed to the dead-house to wait *post-mortem* next morning. At two a.m. the sentry on the dead-house came rushing down to the dispensary (about four hundred and fifty yards off) declaring that he had seen and heard a ghost in the dead-house, to which myself and the compounder and dresser on duty at once proceeded, to find that Clements Twining, who was now partially conscious, was lying on the dead-house flags groaning most piteously—he had rolled off the table on to the floor. He returned to health, and in 1877 accompanied his regiment to England, where I met him at Woolwich in 1883, and he asked me to corroborate his story of 'returning to life' to certain of his acquaintances who had refused to believe him."

"CHOLERA CORPSES REVIVED IN A MORTUARY.

"When the East Norfolk regiment was out cholera-dodging in 1878, Colour-Sergeant T. Hall and Corporal W. Bellomy were sent into cantonments for burial as cholera corpses in the Nowshera Cemetery. There was some delay in the interment owing to a difficulty in obtaining the wood necessary for their coffins, so both bodies were placed in the dead-house, which was generously sprinkled with disinfectants to ward off the risk of contagion. First Hall and then Bellomy regained consciousness, and were duly returned to duty. The following year Bellomy was 'invalided' to England, where I understand he now enjoys the best of health."

"Shortly after the Afghan war of 1878, Surgeon-Major T. Barnwell and I were told off to take a large number of time-expired men, invalids, and wounded, to Deolali on their way to England. Some of the wounded were in a very critical state, necessitating great care; one man in particular, Trooper Holmes of the 10th Hussars, who had an ugly bullet-wound running along his left thigh and under the groin. Our only means of transport for these poor fellows was the 'palki' or doolie carried by four bearers at a curious swinging pace. When we got to Nowshera, Holmes seemed on a fair way to recovery, but the swinging of the doolie seemed too much for him, and he grew weaker day by day till we got to Hassan Abdool, when we could not rouse him to take some nourishment before starting on the march, and to all appearance he seemed perfectly dead; but, as there was neither the time nor convenience to hold a *post-mortem*, we carried the body on to 'John Nicholson,' where, the same difficulties being in the way, and no facilities for burial, we were obliged to put the *post-mortem* off for another day, and convey the corpse to Rawal Pindi rest camp, where we laid him on the floor of the mortuary tent and covered him over with a tarpaulin. This was his salvation, as next morning (*i.e.*, the third day succeeding his 'death'), when we raised the tarpaulin to hold the *post-mortem*, some hundreds of field mice (these tracts are *noted* for them) rushed out, and we noticed that Holmes was breathing, though very slowly—five or six respirations to the minute—and there were a few teeth marks where the mice had attacked his calves. To prevent a relapse by the jolting on further marches, we handed him over to the station hospital staff, who pulled him round, and then forwarded him to the headquarters of his regiment at Meerut."

A lady, distinguished alike for her literary gifts as well as for her philanthropy, sends me the following :—

"I am much obliged to you for sending me 'Perils.' It is a terrible subject, and one that has haunted me all my life, insomuch that I have never made a will without inserting a clause requiring my throat to be cut before I am put underground. Of course one can have no reliance on doctors whatever, and I have myself known a case in which a very eminent one insisted on a coffin being screwed down because the corpse looked so life-like and full of colour that the friends could not help indulging in hopes.

"My great grandmother, after whom I am called, a famous heiress, was a notable case of narrow escape. As a girl she passed into a state of apparent death, and a great funeral was ordered for her. Among the guests came a young girl friend, who insisted that she was not dead, and raised such a stir that the funeral was postponed, and time was allowed to pass till the marvel became that there were no signs of change. I could never ascertain how long this comatose state lasted before she recovered ; but she *did* recover, so thoroughly that after her marriage with Richard Trench, of Garbuly, she became the mother of twenty-two children. Obviously this was no case of a feeble, hysterical, cataleptic subject. I will enclose photograph taken from a miniature of her in a ring in my possession.

"There was another case, well known in Ireland in my youth, of a Colonel Howard, who had a fine place (I think it was called Castle Howard) in Wicklow. He was supposed to be dead, and a lead coffin was actually made with his name and date of death on it ; after which Colonel Howard came to life, and had the plate of the coffin fixed over his kitchen chimney as a warning to his servants not to bury people in a hurry."

Dr. Colin S. Valentine, LL.D., Principal of the Medical Missionary Training College, Agra, N.W.P., told the author during his visit to Agra, February, 1896, that Captain Young, an officer in the regiment of which he (Dr. Valentine) was at that time army surgeon, who had been dreadfully mauled while tiger-hunting in Madras, was laid out for dead, and all the arrangements were made for his funeral at six o'clock that evening, when consciousness returned, and he lived for twenty years after.

In a lecture on "Signs of Death and Disposal of the Dead," delivered by Dr. A. Stephenson at Nottingham, January 9, 1896, the lecturer said "he once attended a

girl living in that locality who was in a trance. All the
preparations were made for her funeral, and the grave
ordered. She remained in a trance three days, and her
mother was annoyed because he would not sign her
death-certificate. On the third day she slowly rose and
recovered. The girl would have been buried unless he
had had a very great fear of her being buried alive."[1]

From the London *Echo*, March 3, 1896.

"NARROW ESCAPE OF A GREEK-ORTHODOX METROPOLITAN.

"A letter from Constantinople, in the *Politische Korrespondenz*,
gives a remarkable case of an apparent death which would have
ended in a premature burial but for the high ecclesiastical position
of the person concerned. On the 3rd of this month, Nicephorus
Glycas, the Greek-Orthodox Metropolitan of Lesbos, an old man
in his eightieth year, after several days of confinement to his bed,
was reported by the physician to be dead. The supposed dead
bishop, in accordance with the rules of the Orthodox Church, was
immediately clothed in his episcopal vestments, and placed upon
the Metropolitan's throne in the great church of Methymni, where
the body was exposed to the devout faithful during the day, and
watched by relays of priests day and night. Crowds streamed
into the church to take a last look at their venerable chief pastor.
On the second night of "the exposition of the corpse," the Metropolitan suddenly started up from his seat and stared round him
with amazement and horror at all the panoply of death amidst
which he had been seated. The priests were not less horrified
when the 'dead' bishop demanded what they were doing with
him? The old man had simply fallen into a death-like lethargy,
which the incompetent doctors had hastily concluded to be death.
He is now as hale and hearty as can well be expected from an
octogenarian. But here it is that the moral comes in. If
Nicephorus Glycas had been a layman he would most certainly
have been buried alive. Fortunately for him the Canon Law of
the Orthodox Church does not allow a bishop to be buried earlier

[1] *Evening News*, Nottingham, January 10, 1896.

than the third day after his death ; whereas a layman, according
to the ancient Eastern custom, is generally buried about twelve
hours after death has been certified. The excitement which has
been aroused by the prelate's startling resurrection may tend to
set men thinking more seriously about the frequent probability
of the cruel horror of the interment of living persons."

The above-mentioned facts have been authenticated
for the author by Dr. Franz Hartmann, of Hallein,
Austria.

NARROW ESCAPES OF SMALL-POX PATIENTS.

Many physicians who dispute the frequency of premature burials admit that the liability to such catastrophes
is considerable during epidemics of small-pox, where
extreme exhaustion, amounting to a suspension of life,
is distinguishable from actual death only by patient and
prolonged observation.

From the *Lancet*, June 21, 1884, p. 1150 :—

"SUSPENDED ANIMATION AFTER SMALL-POX.

" Sir,—I send you privately names and addresses by means of
which you can test, if you please, the accuracy of the following
statements, which I forward for insertion in your journal :—

" Some years since, a young man who had been attacked by
small-pox was declared by the medical man to be dead, and was
laid out for burial. The nurse, however, on paying a visit to the
supposed corpse, thinking there was something uncorpse-like about
its appearance, put a wine-glass over the mouth, and returning in a
quarter of an hour, found it dimmed with breath. He was resuscitated, and, so far as I am aware, is still living. He would now be
about forty-five. He is a farmer.

" A mother and her baby were ill of small-pox, and seemed
likely to die. The grandmother, however, made the nurse promise
that if death appeared to ensue, and even if the medical man
pronounced either or both to be dead, she would put additional

blankets on the one or both, and leave them so till her (the grandmother's) return, which would not be till the next day. They both
appeared to die, and were declared dead by the doctor ; but the
nurse did as she had promised, and the next day when the grandmother returned, they were both alive, and were both living not
very long since.

"Some twenty years ago, I was told that about forty years
previously a young man, in a parish where I was acquainted, was
put in a coffin as a person dead of small-pox ; but when the bell
was tolling for his funeral, and he was about to be 'screwed down,'
he got up and vacated the coffin, and lived several years afterwards.

" In a town where I was brought up, a woman was nearly buried
alive through having gone into a trance on being frightened by a
young lady who had put on a white sheet and pretended to be a
'ghost.' For years she was liable to long spells of insensibility,
from which nothing could rouse her.

"The haste with which small-pox corpses are disposed of
nowadays is to be deprecated. They are usually buried within
twelve hours of their supposed death, and the cases I first
mentioned show with what very probable results. The only sure
proof of death is decomposition, and a law ought to be passed
forbidding burial until signs of it have appeared. Not very long
since I was in a churchyard where a drain was being made round
the church, and was not a little struck by the horrified look of a
labourer who came to the vicar and stated that they had come on
a skull face downward, which, he said, put it beyond doubt that
the person it had belonged to had turned in his coffin after burial.
—I am, Sir, yours faithfully,

"B. A.

"June 18, 1884."

The *Undertakers' Journal*, May 22, 1895, has the
following :—

"The Reverend Harry Jones, in his reminiscences, and as a
London clergyman, declares his conviction that in times of panic
from fatal epidemics it is not unlikely that some people are buried
alive. Mr. Jones recalls a case within his knowledge of a young

woman pronounced to be dead from cholera, and actually laid out for the usual collecting cart to call from the undertakers, when a neighbour happened to come in and lament over her. The story continues thus : 'And is poor Sarah really dead?' she cried. 'Well,' said her mother, 'she is, and she will soon be fetched away ; but if you can do anything you may do it.' Acting on this permission the practical neighbour set about rubbing Sarah profusely with mustard. Sarah sat up, stung into renovated life, and so far recovered as to marry ; 'and I myself,' says Mr. Jones, 'christened four or five of her children in the course of the next few years.' In another case, within Mr. Jones' parochial experiences in London, a man employed as potman lay *in extremis*. A doctor was called in, who said 'Turn him on his face, and I will put a thick strip of flannel soaked in spirits of wine down his spine. We will see what that will do.' A sister brought a store of flannel, the doctor soaked it in spirit, and prepared to apply it as he proposed. First, however, he placed the soaking mass in a heap (almost as big as a small hassock) in the middle of his back. Meanwhile the sister leant forward with a candle and accidently set the hassock on fire. 'This,' adds the anecdotist, 'woke the potman up;' and not very long ago the doctor told me he had seen him in a street near the Oxford Circus."

From the *Daily Chronicle*, September 19, 1895.

" Sir,—I infer from the following facts that numbers of persons are buried alive after being supposed to have succumbed to smallpox.

" Some years ago, at St. Paul's, Belchamp, near Clare, a young man who had been down with the small-pox was pronounced to be dead, and was put into a coffin, which, fortunately, was left unclosed until after the bell began to toll for his funeral, when he rose and stepped out. He lived for many years after. In the same neighbourhood no less than three other similar cases occurred, saving that the undertakers were not so far forward in their work. Each of these would have been buried alive but for the facts that in one case the nurse, having suspicions, put a wine-glass over the mouth of the person (who had been already 'laid out'), and on returning in a quarter of an hour found it dimmed with breath ;

and that in the other case the mother of a mother, who with her baby was declared by the doctor to be dead, had blankets heaped on them, and after a while had the satisfaction of seeing them revive. Two of these three persons are, I believe, still living, and would be just past middle-age. I enclose their names for your private perusal, that you may verify my statements if desired. The first-mentioned case happened about seventy years ago, but I heard of it from residents in the neighbourhood about forty years after it occurred.

"Nowadays as soon as a small-pox patient is supposed to be dead, he or she is enclosed in a coffin and hurried off to the churchyard or cemetery the ensuing night—at least this is the practice in country places. I have no doubt that many have been buried alive.—Yours faithfully,

"EX-CURATE.

"September 18."

Brigade-Surgeon W. Curran cites from the *Revue des Deux Mondes*, April, 1873, in his Eighth Paper, entitled "Buried Alive," as follows :—

"On the 15th of October, 1842, a farmer who lived in the suburbs of Neufchâtel (Lower Seine) went to sleep in his hay-loft in the midst of some newly mown hay. As he did not get up at the usual hour the next morning, his wife went to call him, and found him dead. When the time for his funeral arrived, some twenty-four or thirty hours subsequently, those who were charged with the burial put the body on a bier, and having placed this on the ladder that communicated between the ground and the loft, they allowed it to slide down. All of a sudden one of the rungs of this ladder gave way, and the bier, falling through, was dashed violently on the pavement below. The shock, which might have been fatal to a live person, proved to be the 'saving clause' of our supposed dead one; and fortunately, too, the attendants had not, as so commonly happens in such contingencies, absconded ; on the contrary, responding without delay to the requirements of the situation, they quickly realised the gravity of the crisis, and, unbinding the shrouds of the farmer, they soon restored him to consciousness and life. He was able, we are further told, to resume his ordinary duties in a few days afterwards."[1]

The *Undertakers' and Funeral Directors' Journal*, January 22, 1889, says :—

"Mr. J. W. Smith, of 158 River Avenue, Alleghany, has just had, for instance, a remarkably narrow escape of prematurely putting his family in mourning, and one which will, we may be sure, be a very disagreeable recollection for him during the rest of his existence. After a visit to the Pittsburg Opera House one night, Mr. Smith was found lying 'stiff and cold' behind the stove in the dining-room, and apparently dead. A superficial examination by Dr. M'Cready confirmed the worst fears of Mrs. Smith, but subsequently the doctor sought carefully for any little spark of life which might lurk unseen, and, very fortunately for Mr. Smith, found it. But, beyond that, nothing could be accomplished ; no effort to restore animation produced the slightest effect. Two other physicians were then summoned ; but neither attempts at bleeding, the use of 'mustard baths,' nor the application of electricity, could rouse

---

[1] *Health*, May 21, 1886, edited by Dr. Andrew Wilson, pp. 120-1. After relating other cases, Surgeon Curran continues :—"I have myself personally seen or heard on the spot of three such cases—cases that in other hands or in other localities might have passed as dead, were they not buried as such accordingly."

Mr. Smith after his visit to the opera. For three weeks he lay insensible, and when he regained consciousness a fever followed. This event, and some others of a similar character which are occasionally heard of, show that the examination of persons apparently dead should always be undertaken by an efficient person, and by no means in a perfunctory manner."

The late Madame Blavatsky was subject to death-like trances, and Dr. Franz Hartmann informs me that she would have been buried alive if Colonel Olcott had not telegraphed to let her have time to awaken.

CHAPTER VI.

FORMALITIES AND THEIR FATAL CONSEQUENCES.

WHENEVER grave-yards have been removed, owing to the rapid expansion of towns, in America, or examined elsewhere, unmistakable evidences of premature burial have been disclosed, as will be seen in this volume; bodies have been found turned upon their faces, the limbs contorted, with hair dishevelled, the clothing torn, the flesh mutilated, and coffins broken by the inmates in their mad endeavour to escape after returning consciousness, to terminate life only in unspeakable mental and physical agonies. It may be said that every grave-yard has its traditions, but the facts are carefully concealed lest they should reach the ears of the relatives, or incriminate the doctors who had with such confidence certified to actual deaths which were only apparent. It is not, however, the custom to remove grave-yards in Europe until all possibility of such discoveries has disappeared. To reopen a grave is to break the seal of domestic grief. There is a widespread belief that where a coffin, with a duly certified corpse,—dead or alive,—has been screwed up, it must not be opened without an authorisation from a magistrate, mayor, or other official, and many people have been suffocated in their coffins while waiting for this formality. Common sense, under the circumstances, seems to be often paralysed.

In England it has been decided, Reg. *v.* Sharpe (1 Dearsley and Bell, 160), to be a misdemeanour to disinter a body without lawful authority, even where the motive of the offender was pious and laudable; and a too rigorous interpretation of this and similar enactments in other countries has led to the suffocation of many unfortunate victims of a mistaken medical diagnosis, whose lives, by prompt interposition, might have been saved.

Köppen, in his work, entitled "Information Relative to Persons who have been Buried Alive," Halle, 1799, dedicated to His Majesty the King of Prussia, Frederick William III., quotes the following amongst a large number of cases of premature burial:—"In D——, the Baroness F—— died of small-pox. She was kept in her house three days, and then put in the family vault. After a time, a noise of knocking was heard in the vault, and the voice of the Baroness was also heard. The authorities were informed; and instead of opening the door with an axe, as could have been done, the key was sent for, which took three or four hours before the messenger returned with it. On opening the vault it was found that the lady was lying on her side, with evidences of having suffered terrible agony."

Struve, in his essay on "Suspended Animation," 1803, p. 71, relates the following:—"A beggar arrived late at night, and almost frozen to death, at a German village, and, observing a school-house open, resolved to sleep there. The next morning, the school-boys found the poor man sitting motionless in the room, and hastened, affrighted, to inform the schoolmaster of what they had seen. The villagers, supposing the beggar to be dead, interred him in the evening. During the night, the watchman heard a knocking in the grave, accompanied by lamentations. He gave information to the bailiff of the village, who declined to listen to his tale. Soon afterwards the watchman returned to the grave, and again heard a hollow noise, interrupted by sighs. He once more hastened to the magistrate, earnestly soliciting him to cause the grave to be opened ; but the latter, being irresolute, delayed this measure till the next morning, when he applied to the sheriff, who lived at a distance from the village, in order to obtain the necessary directions. He was, however, obliged to wait some time before an interview took place. The more judicious sheriff severely censured the magistrate for not having opened the grave on the information from the watchman, and desired him to return and cause it to be opened without delay. On his arrival, the grave was immediately opened ; but, just Heaven ! what a sight ! The poor, wretched man, after having recovered in the grave, had expired for want of air. In his anguish and desperation he had torn the flesh from his arms. All the spectators were struck with horror at this dreadful scene."

The *Undertakers' Journal*, November 22, 1880, relates the following :—

"An extraordinary story is reported from Tredegar, South Wales. A man was buried at Cefn Golan Cemetery, and it is alleged that some of those who took part in carrying the body to the burial-ground heard knocking inside the coffin. No notice was taken of the affair at the time, but it has now come up again, and the rumour has caused a painful sensation throughout the district. It is stated that application has been made to the Home Secretary for permission to exhume the body."

Dr. Franz Hartmann, in his " Premature Burial," pp. 10 and 44, relates the two following cases:—" In the year 1856 a man died in an Hungarian village. It is customary there to dig the graves in rows. As the grave-digger was making the new grave he heard sounds as of knocking proceeding from a grave where a man had been buried a few days previously. Terrified, he went to the priest, and with the priest to the police. At last permission was granted to open the grave ; but by that time its occupant had died in reality. The fact that he had been buried alive was made evident by the condition of the body, and by the wounds which the man had inflicted upon himself by biting his shoulders and arms.

" In a small town in Prussia, an undertaker, living within the limits of the cemetery, heard during the night cries proceeding from within a grave in which a person had been buried on the previous day. Not daring to interfere without permission, he went to the police and reported the matter. When, after a great deal of delay, the required formalities were fulfilled and permission granted to open the grave, it was found that the man had been buried alive, but that he was now dead. His body, which had been cold at the time of the funeral, was now warm and bleeding from many wounds, where he had skinned his hands and head in his struggles to free himself before suffocation made an end to his misery."

A medical correspondent communicates to the author particulars of the following case, which occurred at Salzburg, Austria :—" Some children were playing in the Luzergasse Cemetery, and their attention was attracted by knocking sounds in a newly-made grave. They informed the grave-digger of it, and he secured permission to open the grave from whence the sounds seemed to come. A man had been buried there at two p.m. that day. The formalities of the permission to open the grave delayed it till seven p.m., when, on opening the coffin, the body was found to be bent completely over forwards, and was frightfully distorted and bleeding from places on the hands and arms, which seemed to have been gnawed by the man's own teeth. The medical experts who were called in to examine the case declared that the man had been buried alive."

From the *Undertakers' and Funeral Directors' Journal*, January 22, 1887.

"Another shocking case of premature burial is reported ; the distressing incident took place at Saumur, in France. A young man suddenly died, at least to all appearance, and his burial was ordered to take place as soon as possible. The *croquemorts*, or undertaker's men, who carried the coffin to the grave, thought they heard a noise like knocking under its lid, yet, being afraid of creating a panic among the people who attended the funeral, they went on with their burden. The coffin was duly placed in the grave, but, as the earth was being thrown upon it, unmistakable sounds of knocking were heard by everybody. The mayor, however, had to be sent for before the coffin could be opened, and some delay occurred in the arrival of that official. When the lid was removed, the horrible discovery was made that the unfortunate inmate had only just died from asphyxia. The conviction is spreading that the terrible French law requiring speedy interment ought to be modified without delay."

Mr. William Harbutt, School of Art, Bath, writes to me, November 27, 1895 :—" The copies of the pamphlet 'The Perils of Premature Burial,' by Professor Alex. Wilder, you kindly sent me are in circulation. Almost every one to whom I mention the subject knows some instances. One, a case at Radstock, twelve miles from Bath, where the bearers at the funeral heard noises inside the coffin, but were afraid to open it without the authority from a magistrate. When it was opened next day the appearance of the body showed that he had been coffined alive, and had had a terrible struggle to escape."

From the *Star*, London, May 13, 1895.

"A WOMAN LOSES HER LIFE THROUGH LEGAL FORMALITIES.

"Paris, May 11.

"A woman who was believed to have died the day before was being buried at Doussard, when the grave-digger, who was engaged in filling up the grave, distinctly heard knocking coming from the coffin. He called a man who was working near, and he came and listened, and heard the knocking also. It was then about nine o'clock in the morning. The knocking continued, and they listened for about half an hour, when it occurred to one of them that they ought to do something, so they went to inform the local authorities. The curé of the village was the first to arrive on the scene; but as no one had any authority to exhume the body the coffin was not taken up. All that was done was to bore some holes in the lid with a drill in such a way as to admit of air. By mid-day all the necessary formalities had been gone through, and it was decided at last to open the coffin. This was done; but whether the unfortunate woman was still alive at this time is doubtful. Some of those present affirm that she was. They state that they saw a little colour come into her cheeks, and the eyes open and shut. One thing is certain—viz.: that when at half-past six in the evening it was finally decided to consult a doctor, the practitioner summoned declared that death had taken place not more than five or six hours before. It was thought that had the coffin been opened directly the sounds were heard the woman's life might have been saved, and she would have been spared hours of indescribable torture and suffering."

The Paris edition of the *New York Herald*, May 14, 1895, says :—

"The case of the woman buried alive at Annecy, in the Haute-Savoie, the other day, has almost found a pendant at Limoges. A woman, belonging to the village of Laterie, died, to all appearance at least, a few days ago. After the body had been placed in a coffin, it was transported to the village church. On the way the bearers heard sounds proceeding from it, and at once sent for the mayor, who ordered it to be opened. The woman was found to be suffering from *eclampsia*, which had been mistaken for death by her relatives."

The following case is instructive in that the victim was exhumed without an order from the Home Secretary, or waiting for any formalities, and was restored to life :—

"BURYING ALIVE.

[From the *Spectator*, October 19, 1895.]

"Sir,—*Apropos* of your article and the correspondence about being buried alive, in the *Spectator* of September 28, the enclosed may interest you. It is an extract which I have copied to-day out of a letter to a neighbour of mine from his brother in Ireland, dated October 6, 1895 :—'About three weeks ago, our kitchen-maid asked leave to go away for two or three days to see her mother, who was dying. She came back again on a Friday or Saturday, saying her mother was dead and buried. On Wednesday she got a letter saying her mother had been dug up, and was alive and getting all right. So she went up to see her, and sure enough there she was "right enough," as G—— says, having got out of her trance, and knowing nothing about being in her grave from Saturday till Tuesday. The only thing she missed was her *rings ;* she could not make out where they had got to. Her daughter, it seems, told the doctor on her way back here that it struck her that her mother had never got stiff after death, and she could not help thinking it was very odd ; and it made her very uncomfortable.

He never said a word : and the kitchen-maid heard nothing until she got the letter saying her mother was back again and alive. Luckily, she did not "come to" until she had been taken out of her coffin. It was a "rum go" altogether. They say exactly the same thing happened to a sister of hers who is now alive and well.'—I am, Sir, etc.,

"PEVERIL TURNBULL."

CHAPTER VII.

PROBABLE CASES OF PREMATURE BURIAL.

THERE is a great and natural reluctance on the part of medical practitioners to admit that they have made mistakes in death-certification, particularly in any one of the various forms of death counterfeits, or suspended animation. It should be noted that amongst the lectures delivered on special occasions, such as the opening of the medical schools, the subjects of trance and the danger of premature burial are conspicuous for their absence ; allusion to these subjects is of rare occurrence, nor does the study of this abstruse branch of medicine, so far as can be ascertained, form part of any medical curriculum. In the bibliography at the end of this volume, extensive as it is, I can hardly refer to a single instance. Dr. Franz Hartmann, whose work on "Buried Alive" has passed through two English and one German edition, informs me that the same reticence is observable in the medical schools of Germany. Many medical men do not believe in death-trance. They declare that they have never seen such a case, and in their judgment when a sick patient ceases to breathe, when volition is suspended, and the stethoscope reveals no signs of cardiac action, the death is real, and the case beyond recovery. The evidence disclosed in this volume is the result of inquiry in many countries.

From the *Medical Times*, London, 1860, vol. i., p. 65.

"A lady entering upon the ninth month of pregnancy died of pneumonia. All the other phenomena of death ensued, except that the colour of the face was unusually life-like. On the fifteenth day from that of death there was not the least cadaveric odour from the corpse, nor had its appearance much altered, and it was only on the sixteenth day that the lips darkened. The temperature of the atmosphere had undergone many changes during the time mentioned, but although there had been frost for a short period, the weather was in general damp and cold."

This lady might not have been dead. The burial laws should have been such as to make it certain that she was dead before interment, by the appearance of general decomposition. The examination of facts collected by well-known physicians at home leads to the conclusion that cases of narrow escapes from premature burial are by no means of rare occurrence. And it must be obvious to the least reflective reader that in countries where burial follows quickly upon supposed death (as in Turkey and France, some parts of Ireland, and throughout India), or where there is no compulsory examination of the dead (as in the United States or the United Kingdom), and amongst people like the Jews (since the Jewish Law enjoins speedy interment), and especially in cases of sudden death (where attempts at resuscitation are rare), the number of premature burials may be considerable.

In the United States, while there is no law, as in France, enforcing burial within a prescribed number of days, it is the custom of civil authorities, under regulations made by the Boards of Health, to compel interments if delayed by reason of doubt as to actual death beyond a few days.

Particulars of the following case were sent me by a physician, January 17, 1894 :—

"WAS SHE ALIVE?

" Mrs. John Emmons, of North Judson, Ind., was taken suddenly ill and apparently died, a week ago. Her husband desired to keep the body for a few days, to make sure of death. It seems that her mother went into a trance for four days, rallied, and lived five years ; also that her grandfather on her mother's side, after having been pronounced dead for six days, awoke, and lived for twenty-three years. Mrs. Emmons's body was kept until Saturday, when, on the demand of the physician and numerous residents, it was interred. During the time between Monday and Saturday the body did not become rigid. Mortification did not set in, and she was laid to rest without waiting for that, the surest of all tests, to take place. Many are of the opinion that the woman has been buried alive."

There are many cases like the above on record, in which, although there is no absolute proof of premature burial, there is strong presumptive evidence of it. The following from *Truth* (London) of May 23, 1895, is an example, and the writer has heard of many others :—

"The other day I gave a story showing the difficulty of obtaining a *post-mortem* examination after a doctor has once certified the cause of death. One of my readers caps it with a gruesome narrative, of which this is the outline : A man lately died in London. The coffin had to be removed by rail, and was to be closed on the fourth day after the death. My informant, taking a last look at the deceased, was struck by the complete absence of all the ordinary signs of death at such a period. In particular, he states that there was no rigidity in any part of the body, and there was a perceptible tinge of colour in the forehead. He went over to the doctor who had attended the deceased, described all the signs that he had observed, and begged the doctor to come and look at the body before the coffin was closed. The doctor absolutely refused, saying that he had given his certificate, and had no doubt as to the man's death. The friend then suggested that he might himself open a vein and see if blood flowed, to which the doctor replied that, if he did so without the authority of the widow, he would be indictable for felony. Whereupon, says my informant, who was only a friend of the family, 'I had to retire baffled, and let matters take their course.' Why on earth he did not take the widow into his confidence, or risk an indictment for felony by opening a vein on his own account, or even summon another doctor, he does not say. I trust that, should any friend of mine see my coffin about to be screwed down under similar circumstances, and find equal cause to doubt whether I am dead, he will summon up courage to stick a pin into me, and chance the consequences. This, however, has nothing to do with the doctor's responsibilities. It would seem that the medico in this case was either so confident in his own opinion as to decline even to walk across the road to investigate the extraordinary symptoms described to him, or else that he preferred the chance of the man being buried alive to the chance of having to admit he had made a mistake. Which alternative is the worst I do not know."

The *Gaulois* (Paris) of May 16, 1894, contains the following :—

"DEATH OR CATALEPSY?

"The funeral of the Comtesse de Jarnac, whose death was reported to have taken place on Saturday, was fixed for tomorrow, but it will probably be postponed. None of the usual signs of dissolution have appeared; the face still retains its colour, and *rigor mortis* has not yet set in. Some hope is even entertained that the Comtesse may be simply in a state of catalepsy, and that the embolus, to which death was attributed, may have lodged in the lungs, not in the heart, in which case it may merely have caused a stoppage of the circulation (*sic*). The body had not been placed in the coffin up to a late hour last night."

STRANGULATION BY A SCARF.

One of the authors was present on May 14, 1894, with a company of ladies and gentlemen gathered at a country mansion in the Austrian Tyrol for afternoon tea, when the conversation turned upon the subject of premature burial. Among other cases related, the host described that of one of his servants, a woman, who went to bed with toothache, a long scarf being wrapped around her face and neck. As she did not appear the following morning, our host entered her room, and found her, as he supposed, strangled to death by the scarf tightly wound about her neck. A doctor was summoned, when he found that the woman was warm and limp, her face soft and coloured as in life; yet, as there was no respiration or perceptible wrist-pulse, nor beating of the heart, he regarded her as dead, and thought it would be proper to bury her. The host had doubts, however, about the case, and, having decided to observe it further, he had the woman removed to an out-house, where she remained three days longer without any change in her appearance or condition in any way. But, as there was considerable impatience felt at the delay of the burial by the people on the estate, the host sent for two doctors to make a final examination of the woman, and decide as to the existence of life or death. The doctors found that no change had taken place— there was softness of the skin, colour in the face, limpness of the muscles, and an unmistakable warmth of the body; but, as there was an absence of apparent respiration and beating of the heart, they decided that the woman was dead, and urged her burial, which was

9

done. They attributed the high temperature to the process of decomposition which they assumed was going on, though there was no odour of putrefaction noticed by anyone.

The probabilities are that this woman was buried alive. And in the present state of medical education on the subject of apparent death and the causes that bring it about, many physicians would have come to a like conclusion; and, as physicians know but little about it, they are not on their guard concerning its dangers.

A number of cases of apparent death that have survived—where there was strangulation from a scarf, as in this case—have been reported. The explanation in such cases is, that the pressure of the scarf around the neck keeps the venous blood from flowing down from the brain through the jugular veins, and the brain, in consequence, becomes saturated with carbonic acid gas from the detained venous blood, and a death-like stupor caused by carbonic acid poisoning ensues. Artificial respiration would, it is believed, restore such persons to consciousness.

A leading West End undertaker, whose letter is before me, writes under date of June 26, 1896, as follows :—" In my experience I have had but one case come under my personal observation where I had real uncertainty as to death being actually present, and that was an instance of the kind in which this calamity is only likely, in my opinion, to occur. A girl who had been to work in Borwick's factory apparently fainted and died, and within a few days the friends buried her. When we came to close the coffin, there was no evidence

of death, and we did not close it without having a doctor
sent for, and receiving his assurance that she was dead.
When reading the fatal cases which have come to light
upon this subject, I must confess to looking back upon
that instance with much fear, and it is but a poor con-
solation to me that the responsibility was not mine, but
the medical man's."

The foregoing cases are recorded because they are
types of a class that nearly every physician, undertaker,
clergyman, or other observer has met with or heard of,
and the probabilities, having regard to the existing con-
fusion and uncertainty of opinion on the signs of death,
are on the side of apparent rather than real death.   On
the other hand, a medical correspondent informs the
author that he is sceptical as to the reported cases of
narrow escapes, as on more than one occasion his efforts
to verify the facts have proved abortive.   It must be
admitted that there are difficulties in the way of such
inquiries.   If the subject of trance, or narrow escape
from burial, is a lady, publicity injures her prospects of
marriage, and, if a young man, his reputation for business stability is endangered or prejudiced, so that this
reticence on the part of relatives is hardly surprising.
Such persons do not like their gruesome and unpleasant
experiences to be talked about.

CHAPTER VIII..

PREDISPOSING CAUSES AND CONDITIONS OF DEATH-COUNTERFEITS.

THOSE who are most subject to the various forms of death-counterfeit are persons whose vocations exhaust the nervous force faster than the natural powers of recuperation, and who resort to narcotics and stimulants to counteract the consequent physical depression. Dr. Alex. Wilder, in his "Perils of Premature Burial," London, E. W. Allen, p. 19, says :—"We exhaust our energies by overwork, by excitement, too much fatigue of the brain, the use of tobacco, and sedatives or anæsthetics, and by habits and practices which hasten the Three Sisters in spinning the fatal thread. Apoplexy, palsy, epilepsy, are likely to prostrate any of us at any moment, and catalepsy, perhaps, is not very far from any of us." Equally, if not even more likely, to be overtaken by these simulacra of death are the poor—the ill-fed, ill-conditioned, and overworked classes.

With regard to the causation of catalepsy, Dr. W. R. Gowers, in Quain's "Dictionary of Medicine," p. 216, says :—"Nervous exhaustion is the common predisponent ; and emotional disturbance, especially religious excitement, or sudden alarm, and blows on the head and back, are frequent immediate causes. It occasionally occurs in the course of mental affections, and especially melancholia, and as an early symptom of epilepsy."

FAINTING FITS.

Dr. James Curry, F.A.S., in his "Observations on Apparent Death," pp. 81, 82, referring to those conditions and diseases which predispose to death-counterfeits, to which women are more liable than men, says :—" The faintings which most require assistance, and to which, therefore, I wish particularly to direct the attention of my readers and the public, are those that take place from loss of blood, violent and long-continued fits of coughing, excessive vomiting or purging, great fatigue or want of food, and likewise after convulsions, and in the advanced stage of low fevers. It is but seldom, however, that any attempt at recovery is made in such cases; and several reasons may be assigned for this, particularly the great resemblance that fainting fits of any duration bear to *actual death*, and the firm belief of the bystanders that the circumstances which preceded were sufficient to destroy life entirely."

The author continues, pp. 106, 107:—" Nervous and highly hysterical females, who are subject to fainting fits, are the most frequent subjects of this kind of apparent death ; in which the person seems in a state very nearly resembling that of hibernating animals, such as the dormouse, bat, toad, frog, etc., which annually become insensible, motionless, and apparently dead, on the setting in of the winter's cold, but spontaneously revive on the returning warmth of spring. Here, by some peculiar and yet unknown circumstance, the vital principle has its action suspended, but neither its existence destroyed, nor its organs injured, so as absolutely to prevent recovery, if not too long neglected."

Dr. Franz Hartmann reports a case which occurred within half a mile of his residence near Hallein, Austria :—"At Oberalm, near Hallein, there died the widow of a Dr. Ettenberger, a lawyer. It was known that she had previously been affected with fits of catalepsy, and therefore all possible means were taken for the purpose of restoring her to life. All, however, were in vain, and her death appeared to be certain. On the third day, just before the hour appointed for the funeral, the family physician, Dr. Leber, bethought himself of trying some fresh experiments on the corpse, when the woman revived. She had been fully conscious all the time, and aware of all the preparations that were made for her funeral, although unable to make it known to others that she was still alive."

Dr. Hartmann says :—"In 1866, in Kronstadt, a young and strong man, Orrendo by name, had a fit and died. He was put into a coffin and deposited in the family vault in a church. Fourteen years afterwards, in 1880, the same vault was opened again for the purpose of admitting another corpse. A horrible sight met those who entered. Orrendo's coffin was empty, and his skeleton lying upon the floor. But the rest of the coffins were also broken open and emptied of their contents. It seemed to show that the man after awakening had burst his coffin open, and, becoming insane, had smashed the others, after which he had been starved to death."—*Premature Burial, p. 7.*

Bouchut, in "Signes de la Mort," p. 40, relates that "A lawyer at Vesoul was subject to fits of fainting, but kept the matter secret, so that the knowledge of it might not spread and interfere with his prospects of marriage; he only spoke confidentially of it to one of his friends. The marriage took place, and he lived for some time in good health, then suddenly fell into one of his fits, and his wife and the doctors, believing him dead, had him placed in a coffin, and got everything ready for the funeral. His friend was absent, but fortunately returned just in time to prevent the burial. The lawyer recovered, and lived for sixteen years after this event."

INTENSE COLD.

M. Charles Londe, in "La Mort Apparente," p. 16, says:—"Intense cold, coincident with privations and fatigue, will produce all the phenomena of apparent death—phenomena susceptible of prolongation during several days without producing actual death, and consequently exposing the individual who could be restored to life to living burial;" and he further maintains it as an indisputable fact that, every day, people are thus interred alive.

Struve, in his essay on "Suspended Animation," p. 140, says:—"In no case whatever is the danger of committing homicide greater than in the treatment of persons who have suffered by severe cold. Their death-like state may deceive our judgment, not only because such persons continue longest apparently dead, but because the want of susceptibility of irritation is in many cases not distinguishable from real death. A man benumbed with cold burnt his feet, and had continued insensible to pain, nor did he feel this sensation till he warmed them at a fire. In this case it is evident that the susceptibility of irritation was destroyed, while vital power remained."

INFLUENZA.

This is a malady that has been enormously rife all over the world during the past few years, and has baffled the efforts of physicians and sanitarians to arrest its progress: it is sometimes accompanied by conditions which can hardly be distinguished from catalepsy.

The *Lancet*, May 31, 1890, page 1215, gives the following :—

"CATALEPSY AS A SEQUELA OF INFLUENZA.

" The neurotic sequelæ of influenza seem engaging more attention abroad than at home, probably from their symptoms being more pronounced than on this side the Channel. 'Nona,' as it is called, if something more than the somnolence succeeding the exhaustion of influenza, has been thought in Upper Italy to have much in common with catalepsy—one case, indeed, amounting to the 'apparent death' of Pacini. This is reported from Como. The patient, Pasquale Ossola by name, had to all appearance died, and a certificate to that effect, after due consultation, was drawn up and signed. Already it wanted but an hour or so to the interment, when the 'corpse' began to move spontaneously and to exhibit signs of returning life. The relatives of the supposed dead man at once called in assistance, and though animation and consciousness, even to recognition, were restored, the resuscitation was not maintained, and the patient died. Fortunately, the funeral had been arranged on the traditional lines, and the faint chance of return to life was not extinguished by cremation."

NARCOTICS.

Referring to the supposed death of a girl, Sarola, aged eleven years, to whom chloroform had been administered in September, 1894, under peculiar circumstances, and the body hurried off to cremation, Dr. Roger S. Chew, of Calcutta, writes :—" That bottle of medicine was charged with having caused the death of little Sarola, who, I firmly believe, was *burned alive* while in a cataleptic condition induced by the hysterical convulsions, and rendered profound by the administration of the chloroform. Surgeon Lieutenant-Colonel Edward Lawrie agrees with me that at least ninety per cent. of the chloroform deaths are preventable if proper measures are adopted to resuscitate the body, and it is quite possible for a chloroform narcotic to be launched into eternity on the funeral pyre or in the suffocating earth. What a mournful vista Sarola's case opens up, and who can say how many hundreds have been similarly disposed of!"—*Communicated to the Author.*

Sir Benjamin Ward Richardson, in "The Absolute Signs and Proofs of Death," in the *Asclepiad*, first quarter, 1889, p. 9, says :—" In the first experiments made in this country with chloral, after the discovery of its effects by Liebriech, we learned that such a deep narcotism could be induced by this narcotic that it might be impossible to say whether an animal under its influence were alive or dead." And referring to cataleptic trance due to shock, he observes, p. 11, " True traumatic catalepsy is equally remarkable, and equally embarrassing. It has been witnessed in the most destructive form after shock by lightning, and it may also have been met with after severe blows and contusions of the head."

CHOLERA.

Dr. Chew, referring to another of the predisposing causes of apparent death, and the danger of premature burial in India, says :—" In the cholera season there is a risk of a soldier being buried alive, as the custom is to get rid of the body as soon as possible, and it is very seldom indeed that a *post-mortem* is held on a cholera corpse. If the case be one of *true* cholera, decomposition sets in before the breath has entirely left the body, and, immediately life is extinct, putrefaction rushes forward so rapidly as to render a mistake impossible ; but in choleraic diarrhœa or the lighter forms of cholera it is possible that coma resultant on extreme collapse may suspend animation so as to simulate real death *without* actual cessation of vital energy, and lead to live sepulture, except where, by some such lucky accident as the burial ground being a long journey off, the funeral is delayed sufficiently to give a chance of recovery. And this same accident may prove a salvation in syncope or coma from shock or protracted illness.

" With the civil population, save in very exceptional cases, there is very little chance of recovery from apparent death, as the time between alleged decease and sepulture is very short indeed ; and unless there are unmistakable signs of trance, syncope, or coma, the victim must die *after he* (or she) *has been buried alive.*"

VARIOUS PREDISPOSING DISEASES.

Living burials take place because the general public are ignorant of the fact that there are many (some thirty) diseases, and some states of the body that cannot be called diseases, as well as a number of incidents and accidents, which produce all the appearances of death so closely as to deceive any one.

Excessive joy or excessive grief will often paralyse the nervous system, including the action of the heart and the respiratory functions, and occasion the appearance of sudden death as well as shocks, blows upon the head, fright, strokes of lightning, violent displays of temper; also certain drugs now in common medical use, such as Indian hemp, atropia, digitalis, tobacco, morphia, and veratrum. According to Dr. Léonce Lénormand, in "Des Inhumations Précipitées," pp. 85-104, the following diseases and conditions not infrequently produce the like symptoms, viz., apoplexy, asphyxia, catalepsy, epilepsy, nervous exhaustion, ecstasy, hæmorrhage, hysteria, lethargy, syncope, tetanus, etc.

Dr. Herbert Mayo in his "Letters on Truths contained in Popular Superstitions," p. 34, remarks "that death-trance belongs to diseases of the nervous system, but in any form of disease, when the body is brought to a certain state of debility, death - trance may supervene."

Dr. Hartmann observes: "The cases in which persons apparently dead have been restored to health by appropriate means are innumerable, and such accounts may be added to without end, as they are of daily occurrence, while it is also self-evident that, if they had not thus been saved, premature burial and death in the coffin would have taken place. But it also often happens that cases of apparent death recover spontaneously, and even after all possible means taken for the restoration of life have failed. This is specially the case in catalepsy, due to nervous exhaustion, which requires no other remedy than sufficient rest for the recuperation of the life-power, which no kind of medicine can supply."

CHAPTER IX.

THE following are some of the facts and experiences which were brought to the author's notice during a visit to India in the early part of 1896.

THE CALCUTTA BURNING GHAT.

On February 9, 1896, I visited the Burning Ghat on the banks of the Ganges, Calcutta, where twenty bodies are reduced to ashes by fire daily. The corpse of an aged Hindu woman had just been brought in on my arrival, death, we were told, having occurred but an hour before. The deputy registrar asked the nearest relative a few questions as to the age, caste, next of kin, cause of death, which were duly recorded in a book kept for that purpose, and, the charges having been paid, the body, which was as supple as in life (and, except for want of volition, bore no visible marks of death), was placed upon the logs, which were alternately crossed over each other, other logs being placed on the top of the body, with straw underneath. The family being poor in this case, no expensive spiced oils, ghee, or sandal wood were used. The pyre having been sprinkled with water from the sacred river, the nearest male relative took a wisp of lighted straw and ran seven times round it, shouting " Ram, Ram, sach hai " (the god Ram is true and great indeed). He then applied the torch, which in a few seconds reached the body, while a Hindu priest recited verses from the Vedas. The process of burning occupied about four hours. Two other bodies, one an adult, and the other a child, were nearly burnt to ashes during my visit. It appears that in India, when the body is motionless, and assumes a death-like appearance, as in trance or catalepsy, no attempt is ever made at resuscitation, no matter how suddenly or unexpectedly the supposed death may occur, nor is there any proper method of examination for the purpose of death certification. Amongst the Hindus death is not considered an evil, but is the gate leading to a better and happier world. Many Hindus when ill are carried by their friends to the banks of the sacred Ganges, where they meet death with much hope, and without fear.

At the General Hospital, Colombo, I was told by Dr. Van Lagenberg that there was absolutely no protection against premature burials for persons subject to trance, as, although according to the law medical certification was obligatory, medical examination was not; the doctor taking the word of the friends as to the fact of death, and certifying accordingly. Early burial (about six hours after death) was the rule. The Mother Superior to the staff of nurses mentioned the case of the venerable Father Vestarani, an aged Catholic priest of Colombo, who was subject to attacks of epilepsy: these were followed by apparent death, and he had several narrow escapes from premature burial. This case was also known to my friend, Mr. Peter de Abrew, of Colombo, and others. The house surgeon, Dr. H. M. Fernando, said that amongst the Moslems burial followed apparent death very quickly, sometimes in an hour.

From Mr. Vira Raghava Chri, of Madras, manager of the *Hindu*, I learned that the Brahmins always burn the dead soon after death occurs. The relatives, if they reside within easy reach, are sent for. The body is washed in cold water, and after two or three hours the religious service begins, which is performed by the priests, and consists of citations from the Vedas having reference to the departure of the soul from the body, and to the lessons the solemn event teaches. These ceremonies generally last for two or three hours, after which the body is taken to be burned. In answer to my inquiries as to what would happen if within that time no sign of decomposition was exhibited, Mr. Chri informed me that under no circumstances would they wait for more than six hours before the body was taken to be burned. He had heard of cases of persons declared to be dead coming to life while being carried to the funeral pyre, when they were restored to and welcomed by their friends. Cases were also known of the corpse sitting up amidst the flames, and being beaten down by those in charge of the funeral. They were believed to be the victims of premature cremation. He thought, however, that such cases were rare amongst his co-religionists.

Mr. Mohan Chunder Roy, M.B., of Benares, said that it was a very difficult matter, even for a medical practitioner, to distinguish the living from the dead, and, where there were no signs of putrefaction, it was his custom to advise the relatives to wait before burial, or before sending the body to the burning ghat, which they were very reluctant to do. When apparent revivals to consciousness occurred on the pyre, the superstitious people believed that it was due to the presence of evil spirits, and the attempt to escape is frustrated by cremators in charge of the burning ghat. This barbarous custom has been repeatedly affirmed to me by intelligent natives as a matter of common notoriety.

One reason why Hindus are hurried to the cremation ground so quickly, and without waiting to see whether the case is one of trance or suspended animation, is that the relatives are not allowed either to eat or drink while the body remains in the house. If a person touches any article in the house of mourning, that article must be washed and purified. After the cremation all the relatives purify themselves by bathing before they are allowed to eat or drink.

Mr. Durga Prasad, editor of the *Harbinger*, Lahore, writes, February 29, 1896:—"I recollect, when about twelve years old, my grandmother, who was held in great esteem for her piety and experience, told me that she was once declared to be dead, and was therefore carried to our crematorium, or burning-place ; but when about to be burnt she came back to life."

Mr. Joseph, assistant secretary at the Public Library and Museum, Colombo, told the author that his father, owing to weakness of the heart, was subject to frequent attacks of trance-like insensibility. They passed away by simple treatment in a few hours, but were sometimes quite alarming. He was afraid, owing to the superstitious fear of death among the ignorant classes in Ceylon, and the terror which keeping a corpse, or a person in a state of catalepsy, where volition had ceased, excited, that many were buried or burned alive, as it was the custom, particularly amongst the Mahomedans, to carry the body away a few hours after death. Signs of decomposition quickly appeared in a tropical climate, but this unequivocal mode of verifying death was not often waited for by Moslems.

SRI SUMANGALA ON SINHALESE BURIALS IN CEYLON.

Sri Sumangala, the venerable High Priest of the Buddhists of Ceylon, and Principal of the College for Buddhist Priests, at an interview the author had with him in January, 1895, stated that among the Sinhalese the chances of burial or cremation of the apparently dead are not frequent. Their customs are such that a corpse is seldom or never removed for burial or cremation before the expiry of twenty-four hours after death is said to have taken place. During that time climatic influence renders signs of decomposition and putrefaction apparent.

Only one case came under the observation of the venerable theologian, which was that of a person bitten by a cobra. The man apparently succumbed, but a native specialist, having arrived at the cemetery just before the burial, examined the case, and said that life was *not extinct*, and saved the man from a premature grave.

The following is from the *British Medical Journal*, April 26, 1884, p. 844 :—

"PREMATURE INTERMENT.

"The *Times of India*, for March 21, has the following story :— On last Friday morning the family of a Goanese, named Manuel, aged seventy years, who had been for the last four months suffering from dysentery, thinking that he was dead, made preparations for his funeral. He was placed in a coffin and taken from his

10

house, at Worlee, to a chapel at Lower Mahim, preparatory to burial. The priest, on putting his hand on the man's chest, found his heart still beating. He was thereupon removed to the Jamsetjee Jejeebhoy Hospital, where he remained in an unconscious state up to a late hour on last Friday night, when he died."

In a communication to the author from Mr. Nasarvariji F. Billimoria, dated March 14, 1896, the writer says that, where cases of premature burning have occurred in India, the relatives are unwilling to have the facts published, and shrink from making them known. Moreover, when members of a family once declared dead have been rejected by their friends in the land of shadows, and have returned to this life, they are believed to bring misfortune with them, and discredit is attached to the families in consequence. Mr. Billimoria says the following cases can be relied upon as authentic :—

" In the year 18—, in the town of B——, a Marwari was taken as dead and carried to the cremation ground. Unfortunately, at that time a superstition was prevalent among all classes of Indians that, if a dead one is brought back to his or her house, a plague would break out in the town. When, therefore, the Marwari survived, instead of bringing him back to the house, or even allowing him to roam elsewhere, he was killed, it is said, by a hatchet, which they were in the habit of carrying with them to break the fuel for the funeral pyre. This had happened in the old Gaekwari days when Governments did not interfere in the superstitious customs of the people."

Fortunately, however, those days are gone, and with them the old superstitions. Some time ago a fisherwoman, after taking a liberal dose of alcoholic drink and opium, was found (apparently) dead by her relatives—low-caste Hindus. No time is lost among the

Hindus, high or low caste, to remove the body to the cremation ground after a man is found dead.

"A bamboo bier was being prepared to carry the fisherwoman to the *Samashán* (cremation ground), upon which the body was laid as usual, and the relatives were to lift it to their shoulders: when, lo ! the woman turned herself on the bier on her side, and, thanks to the good sense of the fishermen, she is still enjoying her life while I am writing.

"A young daughter of a Bania was sick for a long time, and was found apparently dead by her relatives, and carried to the *Samashán*. These grounds are generally situated at a river side. When the bier was prepared for certain ceremonies, the girl showed signs of revival, and, one by one, the relatives would go near the bier, bend down, stare at the face, and retire aghast. Information had reached the town that the girl had survived ; but the body, nevertheless, was cremated, and never brought back to the house. It is believed that in this case, although the girl had revived for a little time, she had died soon afterwards, as she had been ill for a long time previously. Granting that it was a case in which the dying became actively conscious a few minutes before real death, it is certain that great and indecent haste was practised by the relatives in pressing on the cremation, as is the usual mode in India."

The *Bombay Guardian*, January 11, 1896, under the head of " The Week's News," announced that—

"A Brahmin went to Poona to attend the National Congress. He was laid up with fever, became dangerously ill, and fell into a trance. His friends, thinking him dead, made the necessary arrangements for the funeral. They took the supposed dead man to the river to be burned, but, just as the funeral procession arrived near the Shane temple, his head and hands were seen moving. The cloth having been removed from his face, he opened his eyes and tried to speak. He was taken home."

This case was reported also in the *Times of India*.

The subject of hasty and premature burials in India might with much profit be introduced at the National Congress. The author believes that thousands of people are annually buried and burned in a state of suspended animation—particularly in places where cholera, small-pox, and other devastating plagues prevail. It is usual, both amongst the Parsees and the Hindus, to begin preparations for the religious ceremonies when the case is considered hopeless.

Dr. Roger S. Chew, of Calcutta, who for some years occupied the position of army surgeon in India, writes to me:—" Though there is every risk of live interment with those classes who bury their dead, this is a risk (save in cases of epidemic or battlefield) the British soldier never runs in India, where the military law requires that a *post-mortem* examination, not earlier than twelve hours after decease, must be held on every soldier who dies from any cause except a highly contagious or infectious disease." In the present unsatisfactory state of the law might not this safeguard be generally adopted ?

THE TOWERS OF SILENCE, BOMBAY.

On Sunday, March 15, 1896, my daughter and I were accompanied to the Towers of Silence, situated on the highest part of Malabar Hill, Bombay, by Mr. Phiroze C. Sethna, a highly accomplished Parsee merchant, to whom we were indebted for many acts of kindness during our sojourn in the city. The position is one of rare beauty, commanding as it does charming panoramic views of Bombay and the surrounding neigh-bourhood, while immediately below are extensive cocoa and other tropical plantations. At the entrance to the towers is a notice-board in English, stating that none but Parsees are admitted. We passed under the porch into the sacred enclosure, and found ourselves in the midst of a lovely garden planted with choice shrubs and trees, and were each presented by the gardener with bouquets of freshly-cut flowers.

The towers are five in number, the smallest having been erected in 1669, all modelled after the same pattern, and are about twenty-five feet high. Inside is a circular platform about three hundred feet in circumference paved with large slabs, and divided into rows of shallow open receptacles in which the bodies are placed. There are three sections—for males, females, and children. We noticed a number of vultures sitting on the adjacent trees, and were informed that, when a funeral is on its way, large numbers congregate upon the coping of the tower, ready to seize the body and devour it the moment it is deposited by the corpse-bearers on the slabs, after the conclusion of the funeral ceremonies. In an hour or less the corpse is completely stripped of its flesh, when the bones are thrown into a well. From a sanitary point of view, the plan is preferable to burying or to cremation, which last, as it is carried out in India, is a slow and tedious process. Vultures have never been known to attack children, or even babies left by their mothers tied for safety to a branch of a tree, and will not, it is said, attack a person only apparently dead, as in a trance or coma.

Another custom amongst the Parsees in the treatment of their dead is to bring a dog to the corpse before it is removed from the house, and another dog on its arrival at the Tower of Silence. This ceremony is known as the Sagdeed. In a pamphlet on the " Funeral Ceremonies of the Parsees," by Ervad Jivanji Jamshedje Mody, B.A., a learned priest of the Parsee cult, with whom the author had the pleasure of an interview, the explanation is that, according to the ancient belief, the spotted dog can discriminate between the really and the apparently dead. Dr. Franz Hartmann and other writers appear also to be of the opinion, which the author considers highly probable, that a dog knows whether his master is really dead or only in a trance; but that a strange dog would be able to discriminate and act as a sentinel to prevent a living person being mistaken for a dead one, is highly improbable.

Having heard of several cases of persons taken to the Towers of Silence who recovered consciousness after being laid within the enclosure, I asked Mr. Jivanji Mody what would happen in such a case, and what means of escape there would be? Mr. Mody replied that within the tower there is a chain hanging from the coping to the floor, by which a person could draw himself up to the top of the structure, and he would then be seen and rescued. In a neatly-constructed model of these towers at the museum, Victoria Gardens, Bombay, no chain is visible. The subject of apparent death, or suspended animation, and how to prevent premature burial, premature cremation, and premature exposure in the Towers of Silence, is beginning to excite interest in some parts of India. Mr. Ardeshar Nowroji, Fort Bombay, student of Zoroastrian literature, is to read a paper on the subject before the Debating Society at Elphinstone College. Mr. Soabjee Dhunjeebhoy

Wadia is also studying literature bearing on the same topic.

Mr. Dadabhoy Nusserwanje, a Bombay Parsee and merchant, residing at Colombo, Ceylon, informed the author, January 28, 1896, that he knew of two cases where his co-religionists had been declared dead, and the bodies prepared for burial (the preparation including the long religious service as prescribed by their formulas), who were only in a trance. This was proved by their having come back to life when placed in the Towers of Silence in Bombay. It appears that any persons officially and religiously given over for dead were formerly not allowed to be restored to their relatives, or to the society to which they belonged, as they were supposed to carry with them, from their dead associates, liability to plagues or ill luck, and they are consequently obliged to migrate to distant parts of the country. My informant said that this superstition was so deeply rooted in the minds of the Parsee people that he did not think a reform was possible.

Cases of persons in a trance, mistaken for dead, are by no means uncommon, as would appear from the following communication from Mr. Nasarvariji F. Billimoria, a Parsee of Bombay, addressed to Dr. Franz Hartmann, and not previously published :—

"Several cases of revival of the apparently dead among the Parsees," writes Mr. Billimoria, "have come to my notice.

"A Parsee, whom I shall call M—— B——, was given up as dead. The body was laid on the ground, and the usual ceremonies were being performed, when, to the surprise of the people surrounding the body, he rose and described some spiritual experience. He died long after this event took place, at a good old age, at Bilimora, a town about eighty miles north of Bombay.

"S——, a girl of about ten years, was also taken as dead in the same town, and, after laying her body on the ground, prayers were being recited by the priests. She rose and said that she had been to some other land, where she saw an old lady who ordered her to go away, as she was not required there just then. She died at a good old age a few months ago.

"A woman in the garb of a Hindu beggar was some time ago in the habit of interviewing Parsee ladies at odd times, viz., at about three or four o'clock in the morning, at the same place, and asking several questions pertaining to religion. It was afterwards found that she was K—— (widow of a Parsee priest), who had apparently died a short time before, and, after revival, had emerged from the Tower of Silence, and, a superstition being prevalent among the people that none should be taken back among us who return from the dead, she dared not unite with the Parsees, and hence led a wanderer's life.

"In Bombay, too, I have heard of some cases of the revival of the apparently dead among the Parsees, the principal of them being a lady of a wealthy family, and a Parsee who afterwards carried on his profession as a physician. The physician was living as a Christian on account of the prejudice among the Parsees before referred to. He was called "Mûtchala Dâktar," *i.e.*, doctor with big moustache.

"Similar cases had also occurred in Surat, where two Parsee women had returned from the Towers of Silence, one of whom lived afterwards as a Sanyasini. What became of the other I cannot say."

"The funeral ceremonies among the Parsees provide that, after the signs of death are manifest, the body be washed with warm water, and laid on a clean sheet; two persons hold the hands of the dead person, joining themselves by a *paivand* of tape. The priests recite certain prayers, after which the body is laid on ground set apart for the purpose in the house. Here it lies for several hours, during which time priests recite alternately certain prayers, while a fire is kept alive with fragrant combustibles near the body. The Nasasàlàrs, or corpse-bearers, arrive at the appointed time, when the fire is taken away, and other manthrà or prayers, which occupy an hour or so, are recited by two priests conjointly, gazing first on the iron bier, and then on the face of the body. A procession is then formed, and the body is carried by the Nasasàlàrs only, the others walking in pairs, joining themselves by holding a handkerchief in their hands, several yards distant from the body. The Towers of Silence are removed from the habitations of mankind, sometimes, miles distant, where, after the arrival of the funeral procession, the last obeisance is performed, and the body is carried into the tower, which is called *Dukhmâh*, the mourners, except the Nasasàlàrs, remaining outside. The procession returns after further prayers. The towers are entirely open from above to allow ample sunlight, and to allow the carrion-birds access to the dead.

"From the foregoing it would appear that, with regard to the disposal of the dead, the Parsee system offers advantages, in respect of the revival of the supposed dead persons, over the European system of burial. After real or supposed death, a fire is kept burning near the body, the heat of which would indirectly assist in resuscitating those in a state of suspended animation.

" If a man dies in the afternoon, his body is not carried to the towers till next day, and in that case the fire is kept alive the whole night near the body, two priests alternately reciting manthrâs. Some time is thus allowed to intervene between the supposed death and the disposal of the body in the Towers of Silence. There, too, the body is not laid without Zoroastrian ceremony. But in the system of disposal itself we see another protection, in that the carrion-birds do not touch the body unless they instinctively find evidence of putrefaction. It is a fact that in not a few cases persons have escaped from the dismal and terrible fate of being laid alive in the Towers of Silence. The system of disposal in the tower may appear to non-Zoroastrians repulsive ; but neither the system of cremation nor burial will give us back those whom they have once devoured. That the Parsees do not allow those who have returned from the Towers of Silence to intermingle among them is another question. This too, however, has attracted the attention of this small community ; and I hear that there is a standing order issued from the trustees of the Parsee Panchayet at Bombay to the Nasasálárs (the corpse-bearers) to the effect that they would be rewarded if they would give information or bring back any body which had been revived after it had been carried to the Towers of Silence."

The Parsee custom of using the dog is suggestive. There are numerous cases on record where a dog, following his master to the grave as one of the mourners, has refused to leave the grave ; and these have been quoted as a proof of the undying love of the master's canine friend. May it not be that dogs are gifted, as believed by the Parsees, with another sense denied to most men—the faculty of discerning between real and apparent death? A medical correspondent relates the following :—

"In Austria, in 1870, a man seemed to be dead, and was placed in a coffin. After the usual three days of watching over the supposed corpse, the funeral was commenced ; and when the coffin was being carried out of the house, it was noticed that the dog which belonged to the supposed defunct became very cross, and manifested great eagerness toward the coffin, and could not be driven away. Finally, as the coffin was about to be placed in the hearse, the dog attacked the bearers so furiously that they dropped it on the ground ; and in the shock the lid was broken off, and the man inside awoke from his lethargic condition, and soon recovered his full consciousness. He was alive and well at last news of him. Dogs might possibly be of use in deciding doubtful cases, where their master was concerned."

Also the following :—

"The postmaster of a village in Moravia 'died' in a fit of epilepsy, and was buried three days afterwards in due form. He had a little pet dog which showed great affection towards him, and after the burial the dog remained upon the man's grave and howled dismally, and would not be driven away. Several times the dog was taken home forcibly, but whenever it could escape it immediately returned. This lasted for a week, and became the talk of the village. About a year afterwards that part of the graveyard had to be removed owing to an enlargement in building the church, and consequently the grave of the postmaster was opened, and the body was found in such a state and position as to leave no doubt that he had been buried alive, had returned to consciousness, and had died in the grave. The physician who had signed the certificate of death went insane on that account, soon after the discovery was made."—*Premature Burial, p. 109, London ed.*

CHAPTER X.

THE DANGER OF HASTY BURIALS.

EARLY burials are advocated and defended by certain writers on sanitary grounds; and there is, no doubt, something to be said for them, provided the body shows unmistakable signs of dissolution; but to impose a general rule upon Englishmen by Parliament, or upon Americans by State Legislature, as has been urged, would add to the existing evil of perfunctory and mistaken diagnosis of death, and greatly increase the number of premature interments. The Romans kept the bodies of the dead a week before burial, lest through haste they should inter them while life remained. Servius, in his commentary on Virgil, tells us—"That on the eighth day they burned the body, and on the ninth put its ashes in the grave." Plato enjoined the bodies of the dead to be kept until the third day, *in order* (as he says) *to be satisfied of the reality of the death.* Quintilian explains why the Romans delayed burials as follows:—"For what purpose do ye imagine that long-delayed interments were invented? Or on what account is it that the mournful pomp of funeral solemnities is always interrupted by sorrowful groans and piercing cries? Why, for no other reason, but because we have seen persons return to life after they were about to be laid in the grave as dead." "For this reason," adds Lancisi, in "De Subita. Mort.," lib. i., cap. 15, "the Legislature has wisely and prudently prohibited the immediate, or the too speedy, interment of all dead persons, and especially of such as have the misfortune to be cut off by a sudden death."

Terilli, a celebrated physician of Venice, in a treatise of the "Causes of Sudden Death," sect. vi., cap. 2, says :—" Since the body is sometimes so deprived of every vital function, and the principle of life reduced so low, that it cannot be distinguished from death, the laws both of natural comparison and revealed religion oblige us to wait a sufficient time for life manifesting itself by the usual signs, peradventure it should not be, as yet, totally extinguished ; and if we should act a contrary part, we may possibly become murderers, by confining to the gloomy regions of the dead those who are actually alive."

Mr. Cooper, surgeon, in his treatise on "The Uncertainty of the Signs of Death," pp. 70, 71, had in his possession the following certificate, written and signed by Mr. Blau, a native of Auvergne, a man of untainted veracity :—" I hereto subscribe, and declare, that fifty-five years ago, happening to reside at Toulouse for the sake of my studies, and going to St. Stephen's Church to hear a sermon, I saw a corpse brought thither for the sake of interment. The ceremony, however, was delayed till the sermon should be over ; but the supposed dead person, being laid in a chapel and attended by all the mourners, about the middle of the sermon discovered manifest signs of life, for which reason he was quickly conveyed back to his own house. From a consideration of circumstances, it is sufficiently obvious that, without the intervention of the sermon, the man had been interred alive."

Between 1780 and 1800 many pamphlets on the subject appeared in Germany and France. Opposite sides were taken, some advocating delay until putrefaction, others urging immediate burial.

In 1788, Marcus Hertz wrote strongly against the prevailing precipitate burials among the Jews. He asked "what motive could justify hasty burials;" and continued :—"The writings of learned men and doctors, of both early times and recent date, describe the dangers of precipitate burial; there is not a town in the world that has not its stories of revivals in the grave."

In 1791, Rev. J. W. C. Wolff, in Germany, published numerous narratives of narrow escapes from the grave.

In 1792, Rev. Johann Moritz Schwager stated that he had preached for twenty years against precipitate burials, and that he had been requested to do so by a number of corporate bodies who had evidence of the danger of hasty interments.

About 1800 great excitement prevailed in Germany on account of some narrow escapes from living burial that happened in high quarters, many books and pamphlets having been issued, and sermons preached by the clergy on the subject. The key-note of all of these was the fallaciousness of the appearances of death, and that none was reliable but decomposition.

About this period Dr. Herachborg, of Königsberg, Prussia, wrote that, for forty years, as a doctor, he had always been disgusted with the practice of hasty burials; and, to show the ignorance of the times, he mentions the case of a woman he kept under observation in bed for three days, when her relations took her out and placed her on the floor, insisting that she was dead..

He resisted her burial, and had her covered with blankets; so that by being kept warm she recovered completely. He insisted that no sign of death could be relied upon.

From the *British Medical Journal*, April 12, 1862, p. 390. "The *Gaz. Méd. d'Orient* tells us that people in Constantinople are, in all probability, not unfrequently buried alive, in consequence of the precipitancy with which their burial is performed. If the person dies during the night, he has some chance of escaping premature sepulture; but if he dies during the day, he is sure to be in his tomb in two hours after he has drawn his last breath. Facts of daily occurrence in this country, we are told, prove that persons who were thought to have died during the night have recovered before morning, and thus, thanks to the intervention of night, have been saved from being interred alive. Other facts of not unfrequent occurrence show that persons have recovered while on their road to the grave. In other rarer cases, again, the cries of the revivified half-buried ones have been heard by the passers-by, and thus saved from a horrible conclusion."

In all countries it is the custom amongst the Jews to bury their dead, and apparently dead, quickly, without taking the slightest steps for restoration, and many are the catastrophes recorded.

"The Report of the Royal Humane Society" of 1802 states :—"At the funeral of a Jewess, one of the bearers thought he heard repeatedly some motion in the coffin, and informed his friends. Medical assistance being obtained, she returned to her home in a few hours completely restored."

From the *British Medical Journal*, March 8, 1879, p. 356.

"SUSPENDED ANIMATION.

" A Jew, aged seventy, who had been ailing for some time, apparently died recently in Lemberg, on a Friday night, after severe convulsions. The deceased having been legally certified, the body was put on a bier, preparatory to the funeral, which had to be deferred, the next day being the Jewish Sabbath. Two pious brethren who had, according to their custom, been spending the night in prayer, watching the dead, were suddenly, on the morning of the Saturday, disturbed from their devotions by strange sounds proceeding from the bier, and, to their dismay, saw the dead man slowly rising, and preparing to descend from it, using at the same time very strong language. Both brethren fled very precipitately ; and one of them has since died from the effects of the fright. It is hoped by the *Wiener Medicinische Zeitung* that this case will make the local government watch the Jewish funerals more carefully, as it is known that the Jews often bury their dead very quickly."

The *Undertakers' Journal*, January 22, 1887, says :—
" The dangers that may arise from premature interment are illustrated by a sensational incident which recently occurred at Trencsin, in Hungary. The wife of the Rabbi of the Jewish Congregation apparently died suddenly without having been previously ill. The night before the funeral the female watcher, sitting in an adjoining room, heard a noise in the chamber of death, and, when, stricken with horror, she ventured to open the door, she found that the seemingly dead woman

had risen from her bier, and had thrown off the shroud by which she was covered. By a fortunate accident the interment had been postponed in consequence of the intervening Sabbath, otherwise a horrible fate would have overtaken the Rabbi's wife."

The *Lancet*, August 23, 1884, vol. ii., p. 329, comments thus :—

"BURYING CHOLERA PATIENTS ALIVE.

"It is not so much undue haste as inexcusable carelessness that must be blamed for the premature burying of persons who are not really dead. Such heedlessness as alone can lead to the commission of this crime is not a shade less black than manslaughter. We speak strongly, because this is a matter in regard to which measures ought to be at once taken to render the horrible act impossible, and to dismiss all fear from the public mind. If it be a fact, as would seem to be indisputable, that during the last few weeks there have been cases—we will not attempt to say how many or how few—of burying alive, a scandal and a horror, wholly unpardonable in the last quarter of the nineteenth century, have to be faced ; and the sooner the full truth is known and rules of safety established the better. Let it be once for all decided that measures shall be taken to ascertain the fact of death before burial. Why not revert to the old practice, and *always* open a vein in the arm after death, or pass a current of electricity through the body before the coffin is finally screwed down? It may be held that these unpleasant resorts are unnecessary. We do not think they are. In any case enough is known of the possibilities of 'suspended ani-

11

mation' to render it unsafe to bury until the evidences of an actual extinction of life are unmistakable ; and, as it is impossible to wait until decomposition sets in in all cases of death from infectious diseases, it would be prudent to adopt what must certainly be the least of evils."

If, as the *Lancet* maintains, it is not possible to wait until the only absolute sign of death is manifest, then, in a large majority of cases, there is no safety, and those who die fatally mutilated by horrible accidents may be considered fortunate. The difficulty, we admit, is of a serious nature, particularly for the poor, and can only be overcome by the erection of mortuaries, as discussed in another chapter. The expedient of applying the electric current, suggested by the *Lancet*, has been proved useless in cases of death-trance, where the patients are impervious to the most violent modes of cutaneous excitation.

The *Jewish World*, September 13, 1895, observes :—
" Cases of trance and of the burial of persons who only seemed to be dead, and of narrow escapes of others from the most terrible of all imaginable fates, are not so uncommon as most people suppose ; and while Jews adhere to the practice of interring their dead within a few hours after the supposed demise, there will always be a risk of such horrible catastrophes happening, even more frequently among us than among the general community. Here is, then, really a matter in which some reform is needed, and that without a day's delay.

" To say nothing of the merely human aspect of this important question, to bury until decomposition has actually set in might possibly be shown to be a violation of Jewish Law. It is now commonly admitted that even expert medical men cannot be absolutely certain of death until some signs of decomposition have shown themselves. Now, so strict is the Jewish Law as regards the risk of destroying life, that it is prohibited to even move or touch a man or woman who is on the point of death, lest we hasten, by a moment, their dissolution. It is, therefore, no less than a violation of the Jewish laws against murder to preserve a custom that involves even the minutest scintilla of risk of premature burial. It is high time that this question was seriously taken up by the Jewish clergy and laity."[1]

In the province of Quebec no interment is permitted within twenty-four hours, and the Jews reconcile themselves to this delay, which, however, is far too brief to ensure safety.

It will be said that the danger referred to is not so imminent in the United Kingdom as in France, Spain, Portugal, or even in the United States, owing to the existence of a more temperate climate, and the longer period allowed for burial. This may be so and yet the danger be considerable. It must be remembered that in the rural districts nothing in the shape of examination to establish the fact of death is practised; while in certain parts of Cornwall, throughout the greater part of agricultural Ireland, amongst the Jews in all cities and towns, as well as those who in all places are certified as dead of cholera, small-pox, and other infectious and epidemic diseases, burial often follows certified death

---

[1] For the antiquity of the Jewish practice of early burial, see note in Appendix.

quite as quickly as in the Continental States before
mentioned. In all the public resorts on the Continent
the hotel-keepers, through an insensate fear of death
and the injury which the possession of "a corpse," dead
or alive, may do to their business, have them coffined
and disposed of, particularly in the night, within a few
hours of their supposed death. Dr. D. de Lignières, in
"Pour ne pas être Enterré Vivant," Paris, 1893, says he
has known of burials under such circumstances six hours
after death. This author says that these scandalous
homicidal acts are of every-day occurrence, and that the
rapacious landlords have no difficulty in obtaining
certificates of death from the accommodating *mort
verificateurs*. Every one who visits the *hôtels des villes
d'eaux, des stations balnéaires*, may verify (he says) the
truth of this statement for himself. In short, these are
willing disciples of the "Latest Decalogue":—

> " Thou shalt not kill ; but need'st not strive
> Officiously to keep alive.'

CHAPTER XI.

MANY of those who are most familiar with the phenomena of life and death, including celebrated physicians, men of science, and clergymen, knowing that all the ordinary signs of death (referred to in another chapter) have, in practice, sometimes proved delusive, have been a prey to the suspicion that a fatal mistake is possible in their own case. They have, therefore, left precise instructions in their wills for various preventives which experience has shown to be necessary, and in some instances a combination of these, so as to make doubly sure that they shall not be subjected, like thousands of human beings, to the unspeakable horrors of being buried alive.

Mr. Horace Welby, in his volume entitled " Mysteries of Life, Death, and Futurity," 1861, under the head of " Premature Interment," p. 114, says :—" How prevalent is the fear of being buried alive may be gathered from the number of instances in which men have requested that, before the last offices are done for them, such wounds or mutilations should be inflicted upon their bodies as would effectually prevent the possibility of an awakening in the tomb. Dr. Dibdin relates that Francis Douce, the antiquary, requested, in his will, that Sir Anthony Carlisle, the surgeon, should sever his head from his body, or take out his heart, to prevent the return of vitality; and his co-residuary legatee, Mr. Kerrick, has also requested the same operation to be performed in the presence of his son."

Bishop Berkeley, Daniel O'Connell, and the late Lord Lytton entertained similar apprehensions. Wilkie Collins had a like fear, for he always left on his dressing-table a letter in which he solemnly enjoined his people that, if he were found dead in the morning, he should at once be carefully examined by a doctor. Hans Christian Andersen had a similar dread, and carried in his pocket a note to the effect that, when the time came, his friends were to make sure that he was really dead before burial. Harriet Martineau left her doctor ten pounds to see that her head was amputated before burial. The dread of being buried alive dictated a clause in the will of the distinguished actress, the late Miss Ada Cavendish, for the severance of the jugular vein; and prompted the late Mr. Edmund Yates to leave similar instructions, with the provision that a fee of twenty guineas should be paid for the operation, which was carried out. Mr. John Rose, of New York, who died in November, 1895, made known his earnest desire that his coffin should not be closed, but laid in the family vault at Roseton, and guarded day and night by two caretakers, who were instructed to watch for signs of re-animation.

The late Lady Burton, widow of Sir Richard Burton, provided that her heart was to be pierced with a needle, and her body to be submitted to a *post-mortem* examination, and afterwards embalmed (not stuffed) by competent experts. Lady Burton, it is said, had been subject to fits of trance on more than one occasion, and was terribly afraid that such an attack might be diagnosed as death.

Those who are most apprehensive of apparent death being mistaken for real death are the clergy and other ministers of religion, and funeral directors—in other words, those who know most about it.

Let anyone introduce the subject when in company, on a suitable occasion, and we shall hear of startling cases sufficient to shake credulity, and to compel us to realise the danger to ourselves, as well as to all other members of the community, under our present loose customs. If this dread of premature burial is not universal, as some writers and authorities aver, it is certainly widely extended ; and the evidence set before our readers will show that it is by no means without foundation.

The *Lancet*, March 17, 1866, says :—" There are many apparently trustworthy stories afloat, both in this country and on the Continent, which favour the belief that premature interment not only does sometimes take place, but is really of not so unfrequent occurrence as might be supposed. Some few believe it to be not an unlikely event, and break out into a cold perspiration at the thought of the possibility of the misfortune happening to themselves. Others have actually made provision in their wills that means should be taken, by cutting off a finger, or making a pectoral incision, etc., to excite sensibility, in case any should remain after their supposed death; whilst a French countess, in order to escape so terrible a fate, left a legacy to her medical attendant as a fee for his severance of the carotid artery in her body before it was committed to the tomb."

The Rev. John Kingston, chaplain R.N., writing to
the (London) *Morning Post*, September 18, 1895, says—
"The danger of being buried alive appears to be a
very real one ; and I can testify, from my experience
as a clergyman, that a great many persons are haunted
by the dread of that unspeakably horrible fate." The
writer further expresses a hope that the ventilation of
the subject will be followed by practical results.

While speaking on the subject of premature burials,
in a lecture delivered at Everett Hall, Brooklyn, New
York, June, 1883, Mr. J. D. Beugless, the then President of the New York Cremation Society, said that an
undertaker in that city (Brooklyn) recently made provision in his will, and exacted a promise from his wife of
great caution, that his body should be cremated, being
induced thereto by the fear of being buried alive. " Live
burials," he says, "are far more frequent than most
people think." It is reported that another undertaker of
Brooklyn some time since deposited a body in a receiving vault temporarily : when he went some days
later to remove it for burial, what was his horror, upon
opening the niche in which the coffin had been placed,
to find the body crouching at the door, stark in death,
the hair dishevelled, the flesh of the arms lacerated
and torn, and the face having the most appalling
expression of horror and despair ever witnessed by
mortal eyes !

An undertaker, writing to the *Plymouth Morning
News*, October 2, 1895, mentions that he reluctantly
buried a young person, who lay in the coffin for seven
clear days without sign of decomposition, and only
consented to close the coffin then, on the assurance that

the same conditions attended all the deaths which had previously occurred in the family. Dr. Hartmann and other authorities have found that such cases are probably the subjects of catalepsy, a malady which sometimes runs in families and affects every member. The undertaker adds that, in future, he should decline to close the coffin of the apparently dead until signs of decomposition set in, "thus preventing the possibility of our worst fears being realised." If undertakers generally would adopt these wise and necessary precautions, living sepulture would come to an end. Under the existing imperfect system of medical examination—and, as we have shown, both in England and in the United States, where there is usually no examination at all—there is often a reckless haste in interments. No thoughtful persons can contemplate the burial of a million and a half human beings annually in these two countries without mistrust and misgivings.

Many well-to-do people in civilised countries provide in their wills for the prevention of premature interment, by leaving instructions for surgical operations after their decease, *post-mortems*, embalmment, or cremation. It may happen, however, that wills are mislaid, lost, or withheld by the testators, or are not opened and read until after the funeral, when the instructions in this regard, however strictly enjoined, are rendered abortive. Legacies should be given conditionally on the observance of certain duties, and only payable on proofs to the executors that they have been carried out. A large majority of people do not, however, leave testamentary instructions, for the simple reason that they have nothing to bequeath. And the majority have an

equal claim with the minority to be safeguarded by the
State against such terrible misfortunes.   Syncope, sometimes mistaken for death, is a condition to which both
men and women, who are compelled by their poverty
in all large cities to endure exhausting labours in ill-
ventilated work - rooms, and their often ill - nourished
children in board schools in England and in the public
schools in America, are peculiarly liable.

CHAPTER XII.

SUDDEN DEATH.

THE idea commonly entertained is that with animal bodies there are only two possible conditions—either life or death; that the presence of one of these conditions implies the absence of the other; that when the body has assumed the appearance of death, as during the sudden suspension of all the functional activities, it must be dead. This last is far from being true; for all the appearances of death are fallacious, especially those that accompany so-called sudden death. All such cases should be challenged as of doubtful character, and held so till recovery or putrefaction of the tissues proves the presence of life or of death. This subject is too often treated by medical writers with indifference. Technically, it is regarded as a failure of the brain, or lungs, or heart, to perform their functions; popularly, we say that "the thread of life is snapped asunder;" or it is "the going out of life," like the sudden extinguishing of a candle. The author's experience, however, at the sick bedside, and in the death-chamber, has taught him that life leaves the body in a gradual manner, and that death approaches, and takes the place of life, in one part or organ after another, thus creeping through the tissues, and sometimes defying all tests to prove its presence, leaving putrefaction to be its only sign. There can be no such thing as veritable sudden death, unless the body is crushed into a shapeless mass, like an insect under foot.

The late Dr. Farr, of the Registrar-General's Department, London, says :—"No definition of the sense in which *sudden death* is practically understood by coroners has been given." Dr. Granville says : "The writers on medical jurisprudence do not state with any strictness what they mean by sudden death, whether it be death in ten minutes, ten hours, or ten days."[1] And he asks in the same vein, "Does sudden death mean death in three minutes, three hours, or three days?"[2] Still further he remarks regarding the customary definitions, "They lead one to infer that a certain mysterious principle, called LIFE, has been instantaneously withdrawn from a healthy and well-constituted individual, who was at the very moment, as heretofore, exercising his proper animal functions with a regularity that promised to endure for a long continuance of years. . . . No such phenomena occur in Nature, unless through violence or from accident. Under Nature's laws there is no such thing as sudden death. . . . In every case where death has abruptly cut short the thread of life, there has been a preparation, more or less antecedent to the occurrence, which must inevitably have led to it. . . . The victim may seem to have been struck down, as if by lightning. But in reality the event was only the natural termination of an inward state of things which insidiously and unexpectedly was preparing the blow."[3]

Dr. Tidy, in "Legal Medicine," p. 29, says :—"As a rule, the action required to bring about complete molecular death—*i.e.*, the suspension of vital activity in every

---

[1] Dr. A. B. Granville, "Sudden Death," p. 278.　　[2] Ibid., p. 278.

[3] Ibid., p. 279.

part—is progressive. In a given case, therefore, we are unable to state any definite time as the period of its occurrence. The popular idea of death is that the entire body dies at once. Somatic death is an impossibility." Thus, it is clear that the process of death, or the departure of life, may require days or weeks for its completion; and it may even be delayed to a time when putrefaction has set in quite generally, as when the hair and nails grow after the body has been buried some weeks, as has been credibly reported. Writers upon so-called sudden death recite a number of diseases and conditions which quickly destroy the machinery that carries on the vital functions, thus rendering resuscitation quite impossible. Tidy[1] names some twelve of such causes: prominent among them are diseases of the heart, rupture of the heart, clots in the blood vessels, aneurisms, effusions of blood in the brain, bursting of visceral abscesses, ulcers of the stomach, extra-uterine pregnancy, rupture of the uterus or bladder, large draughts of cold water taken when the body is heated, cholera, alcoholic poisoning, mental emotions, etc. But he remarks upon these causes—" Because a person dies suddenly, there being no evidence of violence or poison, the action adopted by many coroners in not requiring a *post-mortem* examination leaves the most important witness—the dead body itself—unheard, and the inquest so far valueless." Which may mean that, without the risk of an autopsy, it is impossible in such cases to determine whether they are beyond resuscitation or not, unless putrefaction settles the question. Unfortunately there is nothing in the

---

[1] Tidy, " Legal Medicine," part i., pp. 279-280.

external appearance of those cases of so-called sudden death in which the vital machinery may be totally wrecked, to distinguish them from those of apparent death, in which all the organism is in a state of perfect integrity, and in which resuscitation is possible, provided the vital principle has not entirely left the body. Consequently, the only safe rule to observe in all cases in which death has not followed poisoning, or injuries which kill outright, or some known disease of sufficient duration and severity to bring on dissolution, is to wait for unmistakable evidences of decomposition before autopsy, embalming, cremation, or burial is allowed.

In former times precipitate interments of persons who died suddenly were specially guarded against.

Nothing is more common, on opening a newspaper, than to see one or more announcements of sudden death. These occurrences are so frequent that the great London dailies, except when an inquest is held, or when the deceased is a person of note, omit to record them. The narratives are much alike : the person, described to be in his usual health, is seized with faintness in the midst of his daily avocation, and he falls down apparently dead ; or he retires for the night, and is found dead in his bed. In many instances *post-mortems* are made, and an inquest held ; but in other cases the opinion of the attendant doctor, that the death is due to heart-disease, syncope, asphyxia, coma, apoplexy, or "natural causes," is deemed sufficient. The friends who are called in to look at the body will remark, " how natural and how life-like," " how flexible the limbs," " how placid the face ; " and, without the faintest attempt at resuscitation, arrangements are made for an early burial.

Dr. Alexander Wilder, Professor of Physiology and Psychology, in a letter to the author, says :—" There are a variety of causes for sudden death. The use of tobacco is one. Another is overtaxed nervous system. Men of business keep on the strain till they drop from sheer exhaustion. At the base of the brain is a little nerve-ganglion, the medulla oblongata, which, once impaired, sends death everywhere. Overtaxing the strength by study and mental stress will do this. The solar ganglion below the diaphragm is the real vital focus of the body. It is first to begin, last to die. A blow on it often kills. An emotion will paralyse it. Even undue excess at a meal, or the use of overmuch alcohol, may produce the effect.

" Tobacco impairs the action of the heart. An overfull stomach paralyses the ganglionic store, and breathing is likely to stop. It is dangerous in such cases to lie on the back. All these deaths are by heart-failure." It is syncope where the heart fails first ; asphyxia where the lungs are first to cease ; coma when the brain is first at fault. " Natural causes " and " heart-failure " usually mean, like " congestion," that the doctor's ideas are vague.

Dr. Wilder continues :—" I would choose such a death if I could be sure it was death. *But most of those things which I have enumerated may cause a death which is only apparent.*"

The following briefly extracted cases from English papers are typical of thousands of others, and can be duplicated, with slight variation in terms, throughout the

United States.   The absolute proof of the reality of
such deaths is not found in hasty diagnosis or in medical
certificates, but in the presence of putrefaction :—

"SUDDEN DEATH AT ST. AUSTELLS.

"Mr. P. G—— died suddenly yesterday.   Apparently in his
ordinary health, he had been busily occupied during the morning ;
went upstairs, and was found lying on his face on the floor.   Dr.
Jeffery was called, and pronounced life extinct, and expressed the
opinion that death arose from syncope."— *Western Morning News*,
September 14, 1895.

"SUDDEN DEATH IN PEASCOD STREET.

"An inquiry was held as to the circumstances attending the
death of W. P——, which took place suddenly the previous
evening.   The deceased was forty-three years of age, and invariably enjoyed good health, except that he complained of
headache at times.   The jury returned a verdict of death from
natural causes."— *Windsor Express, September 21, 1895.*

"SUDDEN DEATH.

"T. B—— was seized with sudden illness after retiring to rest,
and expired before medical aid could be obtained.   Deceased
had been in his accustomed health, had been at work all day, and
had eaten a hearty supper before retiring to rest.   The Coroner
was communicated with ; but, as death was certified to be due to
heart-disease, no inquest was necessary." — *Middlesex County
Times, October 2, 1895.*

"SUDDEN DEATH OF A SERVANT.

"The deceased, L. E——, aged twenty, retired on Sunday
evening in her usual state of good health.   In the morning she was
found insensible, and when the doctor arrived, shortly afterwards,
he found life to be extinct.   Evidence was given to show that she
had previously been perfectly bright, cheerful, and well.   Verdict
of the jury, that ' Deceased died from failure of the action of the
heart in the natural way.' "— *Harrogate Advertiser, October 12,
1895.*

"AWFULLY SUDDEN DEATH NEAR AMBLESIDE.

" Mr. H——, who had been remarkably cheerful during the day, was just in the act of lighting his pipe to enjoy a smoke,when his head fell back, and he died in a moment. The family doctor certified to the cause of death."—*Lancaster Guardian, October 12, 1895.*

"SUDDEN DEATH AT SEA.

" Mr. R. B. Tobins, the County Coroner, held an inquiry at the Guildhall, Plymouth, concerning the sudden death of P. E——. The deceased was sixty years of age, and was speaking to William Parkinson, when he began to cough, and passed away suddenly. Witness never knew deceased to be ill. Dr. Williams made a superficial examination of the body, and attributed death to heart-disease. Verdict : ' Natural causes.'"—*The Western Mercury, Plymouth, October 22, 1895.*

"SUDDEN DEATH AT TWICKENHAM.

" Lieutenant S. C. G—— fell down and expired suddenly while walking near Kneller Hall, yesterday afternoon. Deceased was forty-four years of age, and had been in his usual health."—*Daily News, November 1, 1895.*

" SUDDEN DEATH AT FOREST OF DEAN.

" Mr. J. W. W—— died very suddenly. He was forty-five years of age ; in his usual health and spirits on Monday ; slept well ; got up at five ; told Mrs. W. W—— he was giddy ; felt ill ; went to bed ; and died in her arms in a few minutes."—*Western Press Bristol, November 1, 1895.*

"SUDDEN DEATH, WESTON-SUPER-MARE.

" Mrs. E. T—— was found dead in her bedroom. She appeared 'all right' when she retired to rest on Monday evening." —*Bristol Times and Mirror, November 7, 1895.*

" SUDDEN DEATH AT NELSON.

" The East Lancashire Coroner has received notice of the death of Ann, the wife of T. B——. She retired to bed apparently all

12

right on Friday night. At two a.m. on Saturday the husband, who was awakened by the crying of the baby, went to his wife's bedroom and found her dead, she having apparently died in her sleep."—*Lancashire Express, Blackburn, November 11, 1895.*

"SUDDEN DEATH.

"A painful shock was caused at Lowestoft last evening by the sudden death of Mr. T. R.——, who was forty-seven years of age, and apparently in his usual health. He drove out to pay a visit, but death took place a few minutes after his arrival."—*Morning Advertiser, November 19, 1895.*

"SUDDEN DEATH AT LLANDERFEL.

"Mr. D. L—— was found dead in bed on Sunday morning at half-past eight. The deceased, who was fifty-four years of age, was apparently in the best of health on Saturday, and had come on a visit to his daughter. The verdict at the inquest was: 'Death from natural causes.'"—*Western Mail, Cardiff, November 19, 1895.*

"SUDDEN DEATH.

"On Tuesday morning, between nine and ten o'clock, A. S——, thirty-six, was in her bedroom apparently in her usual health, when she suddenly fell back against a chair and expired."—*Portsmouth Mail, November 28, 1895.*

"DIED AT HIS WORK.

"Yesterday the district Coroner was notified of the death of T. C. F——, aged thirty-nine, a butcher. F—— was cutting some meat on the block when he suddenly fell backwards dead. He had always enjoyed excellent health."—*Sun, November 29, 1895.*

"SUDDEN DEATH AT EAST GRINSTEAD.

"Mr. W P——, a carpenter, died suddenly yesterday morning. He was engaged at a light task at his bench, apparently in his usual health, when about ten o'clock he was seen to fall backwards. The doctor on arriving could only pronounce life extinct." —*Sussex Daily News, December 4, 1895.*

"SUDDEN DEATH.

"W. D. D—— died suddenly yesterday morning. Deceased appeared to be in his usual health when he retired on Monday. About half-past six in the morning he was supplied with a cup of tea, and an hour later was found dead in bed. Dr. R—— was called in, and said death was due to natural causes."—*Dundee Advertiser, December 4, 1895.*

"SUDDEN DEATH OF A TRAM CONDUCTOR ON DUTY.

"A shock was occasioned the passengers as they were proceeding to town this morning by the sudden death of the conductor in charge. The deceased, J. D——, whose age is twenty-nine, had always been a steady, faithful servant, an army reserve man, and *suffered from no ailment, and certainly not from one likely to cause sudden death.*"—*Daily Argus, Birmingham, December 5, 1895.*

"SUDDEN DEATH OF A WALSALL LABOURER.

"On Tuesday, E. W——, aged thirty-six, retired to bed to all appearances in his usual health. His wife tried to awaken him about a quarter past seven on the following morning, but found that her husband was dead."—*Wolverhampton Evening News, December 6, 1895.*

"SUDDEN DEATH OF A COLLIERY MANAGER.

"Last night Mr. A. B. Stouth held an inquest concerning the death of T. S——. The deceased, who was described *as a very healthy man*, went to the colliery shortly after six o'clock; he conversed freely with the workmen, and when in the act of taking off his coat he fell down and died. The verdict, without *post-mortem*, was returned: 'Died from natural causes.'"—*Birmingham Daily Gazette, December 10, 1895.*

"SUDDEN DEATH OF A VICAR.

"The Rev. T. S. C——, of Salop, died very suddenly at his residence. He attended to his usual duties in the morning, apparently in the full enjoyment of health, and in the afternoon conducted a funeral. Immediately upon his return he was taken ill, and died a few minutes afterwards."—*Daily Argus, Birmingham, December 16, 1895.*

"SUDDEN DEATH OF A RAILWAY EMPLOYEE.

"A painfully sudden death occurred at Hounslow. A. H——, aged nineteen, clerk, started from home to attend his duties at the office, apparently in robust health. At about eight o'clock, whilst sitting between two companions at a table, he suddenly fell forward and expired."—*Hounslow Chronicle, December 21, 1895.*

"SUDDEN DEATH OF A SCHOOL BOARD MEMBER.

"A painful sensation was created at Leicester yesterday by the discovery that Mr. R. M——, a leading Wesleyan, had been found dead in his bed. He was apparently in excellent health when he retired, after a light supper."—*Middlesborough Daily Gazette, December 30, 1895.*

"SUDDEN DEATH.

"Major Taylor held an inquest on C. N. W—— yesterday. The deceased was described as a fine healthy boy. On Sunday forenoon he was placed on his grandmother's knee to nurse, when he fell back and expired. A verdict of death from natural causes was returned."—*Evening Press, York, January 1, 1896.*

"SUDDEN DEATH.

"Yesterday, Mr. Reilly, Coroner, held an inquest on H. A. C——. It appeared that the servant, in passing his room, heard him moaning. Medical aid was procured, but he died in a few minutes. Deceased was in the enjoyment of robust health previously. Verdict: 'Death from natural causes.' "—*Irish Times, Dublin, January 3, 1896.*

"SUDDEN DEATH.

"Mr. H. W—— was suddenly taken ill between five and six yesterday evening, apparently suffering from an apoplectic fit, and expired in a few minutes. Mr. W—— was a gentleman enjoying most robust health, and earlier in the afternoon was chatting genially with several of his friends. An inquest will probably not be necessary."—*Darlington North Star, January 17, 1896.*

"SUDDEN DEATH.

"The City Coroner held an inquiry on Saturday at the Stanley Arms relative to the death of Alice M. A——, aged twenty-eight, who died suddenly. On Friday she seemed in good health and spirits. From an internal examination of the body Dr. Miller was of opinion that she died of syncope or failure of the heart's action. Verdict: 'Death from natural causes; to wit, heart-disease.'"—*Eastern Daily Press, Norwich, January 20, 1896.*

Amongst other sudden deaths more recently reported are :—R. F——, of Torquay, described as "a man of exceptional physique, who had every appearance of possessing a very robust constitution."—F. P. C—— "looked more than usually robust of late, had never been known to complain of his head, and appeared in the best of health and spirits."—W. W—— "had always appeared to enjoy good health, with the exception of a cough."—O. P——, "beyond failing appetite, had given no indication of ill health."—W. M—— "was in his usual health, and went to bed all right."—Mrs. T. B—— "was in the best of health, and was attending to her household duties."—L. T——, "a powerfully-built fisher-man, and most unlikely to come to such a sudden termination of life."—M. J. M——, at East Garston. "A *post-mortem* was made by Dr. K. and his assistant, but they were unable to find any evidence as to the cause of death. Verdict: 'Natural causes.'"—The sudden death, while playing the pianoforte, of a girl, aged twelve, "who had never had a day's illness in her life." — S. G—— "was quite well, and in excellent spirits."—T. B. B—— was "a robust man, and had not been ailing."—G. R—— was "in excellent health and spirits, and attended to his duties as usual."—A little

girl, M. B——, who appeared to be in her usual health,
died very suddenly while sleeping in a cot by the side
of her parents. Verdict at the inquest: "Death from
natural causes."—A. S——, aged twenty-three, a strong
young fellow, who went to rest before eleven o'clock.
About one o'clock the following morning he was seized
with pain, became unconscious, from which he succumbed.—R. J. C——, labourer, "a fine, robust-looking
man," suddenly expired before medical aid could be
procured. Verdict at inquest: "Died suddenly from
natural causes."—Mrs. R——, "who was quite well when
her daughter left the room, was found dead on her return
a few minutes later."—T. H——, blacksmith, "went to
bed in his usual health and spirits" in company with a
comrade, who on attempting to wake him in the morning found life extinct.

The above are given simply as typical examples of
a class of cases of which thousands might be cited, but
it has not been thought necessary to weary the reader
with the details of further instances.

While it is not suggested that the foregoing are cases
of premature burial, yet it is absolutely certain that they
belong to the category of persons of whom a considerable
percentage are liable to such misadventures unless precautions very different from those in vogue are taken to
prevent them. All medical practitioners allow that a
man may be half drowned or half dead, and that cases
of suspended animation occur where the most experienced
physician is unable to detect the faintest indication of
breathing or cardiac movement. They are, however,
quite sceptical as to absolute suspensions of life where
all the ordinary methods to test its existence fail ; and,

owing to this scepticism, and the readiness to give certificates of death in cases of alleged sudden death, have unwittingly promoted premature burials, as will appear by the facts quoted in these pages.

Mr. M. Cooper, in the "Uncertainty of the Signs of Death," p. 49, cites from a letter by one William Fabri, a surgeon, the opinion that we " . . . have just reason to condemn the too precipitate interment of persons overpowered by lethargies, apoplexies, or suffocation of the matrix; for I know there have been some, supposed to be irretrievably cut off by these disorders, who, resuming strength and returning to life, have raised the boards of their own coffins, because in such disorders the soul only retires, as it were, to her most secret and concealed residence, in order to make the body afterwards sensible that she had not entirely forsaken it." These wise counsels were written two hundred and sixty-eight years ago, since which time thousands of our fellow-creatures have, it is feared, been the victims of premature interment, and yet the danger then pointed out remains. The *Undertakers' and Funeral Directors' Journal*, the conductors of which are laudably anxious to keep their profession from the odium of burying people alive, referring to sudden deaths and this danger, says, in its issue of January 24, 1894, under the head of "A Burning Question":—"Sufferers from such chronic ailments as are reputed to end suddenly are in constant danger from the present state of the law, if they are in the hands of people interested in their death." And continues: "Even where a medical certificate is obtained, such general laxity has entered into proceedings that but little protection is thereby afforded to the public. While the medical man is bound to state what he believes to be the cause of death, he is under no obligation to make sure either that the patient is dead at all, or that, if dead, he died from a particular disease for which he was attending him."

VIVISECTION.

The *Medical Times and Gazette*, 1859, vol. xviii., p. 256, has the following :—

"A CRIMINAL'S HEART.

" We find in an account taken from the 'Boston Medical and Surgical Journal' some observations on the heart of a hanged criminal, which are remarkable in a moral point of view, as well as in their scientific aspect. The man died, it appears, as the phrase is, without a struggle ; and, therefore, probably in the first instance, he fell into a syncope. The lungs and brain were found normal. Seven minutes after suspension, the heart's sounds were distinctly heard, its pulsations being one hundred a minute; two minutes later they were ninety-eight ; and in three minutes sixty, and very feeble. In two minutes more the sounds became inaudible. The man was suspended at ten o'clock, and his body was cut down twenty-five minutes afterwards. There was then neither sound nor impulse. At 10.40 the cord was relaxed, and then the face became gradually pale ; the spinal cord was uninjured. . . . At 11.30 a regular movement of pulsation was observed in the right subclavian vein ; and on applying the ear to the chest, there was heard a regular, distinct, and single beat, accompanied with a slight impulse. Hereupon Drs. Clark, Ellis, and Shaw open the thorax, and expose the heart, which still continues to beat ! The right auricle contracted and dilated with energy and regularity. At twelve o'clock the pulsations were forty in a minute ; at 1.45 five per minute. They ceased at 2.45 ; but irritability did not entirely disappear until 3.18, more than five hours after suspension. 'This fact,' says M. Séquard, 'demonstrates that in a man, unfortunately, even when syncope exists for some minutes at the commencement of strangulation, the ventricles of the heart cease to beat almost as quickly as they do in strangulation without syncope.' With regard to the moral aspects of this case, the same gentleman remarks :—'People will probably be surprised that the body of this man should have been opened while the beating of the heart was still audible. We will not ask here if the doctors committed or not a blamable action ; we will only say that we know them personally, and that, if they have in part merited the violent reproaches addressed to them, they are, nevertheless, *hommes de cœur,* who, in an excess of scientific zeal, did not notice that the body upon which they experimented was not, perhaps, at the time a dead body.'"

SYNCOPE.

The deaths attributed to syncope in the Registrar-General's reports for England and Wales during the last six years are :—

	MALES.		FEMALES.
1888	....	817	....	896
1889	....	939	....	922
1890	....	1,237	....	1,250
1891	....	1,355	....	1,301
1892	....	941	....	943
1893	....	848	....	770

Syncope, however, is not a disease, though often certified as such, but is merely a symptom of certain maladies, or a manifestation of suspended animation from unascertained cause. In Hoblyn's " Dictionary of Medical Terms," p. 632, syncope is described as—" Fainting or swoon; a sudden suspension of the heart's action, accompanied by cessation of the functions of the organs of respiration, internal and external sensation, and voluntary motion." There appears, therefore, every probability that, with careless or ignorant medical practitioners, syncope is not seldom mistaken for trance, and a certificate of death may be given where there is merely a suspension and not a termination of life ; and this probability is reduced to a certainty when we learn the number of premature burials and narrow escapes reported by Winslow, Bruhier, Köppen, E. Bouchut, Lénormand, F. Kempner, Moore Russell Fletcher, Gannal, Gaubert, Hartmann, and other recognised authorities. Dr. James Curry, Senior Physician to Guy's Hospital, and Lecturer on the Theory and Practice of Medicine, in the introduction to his " Observations on Apparent Death," London, 1815, 2 ed., p. 1, says:—"The time is still within the recollection of many now living when it was almost universally believed that *life* quitted the body in a very few minutes after the person had ceased to breathe. Remarkable examples to the contrary were, indeed, upon record ; but these, besides being extremely rare, were generally cases wherein the *suspension*, as well as the *recovery of life*, had occurred *spontaneously;* they were, therefore, beheld with astonishment, as particular instances of Divine Interposition." It is believed that the majority of the members of the medical profession still entertain the idea that a human being is dead when breathing can no longer be detected, as in the cases of reported sudden deaths ; and, except in those which occur from drowning, or suffocation through noxious gases, attempts are very rarely made to promote restoration, and, unless they return to life spontaneously while above ground, there are good reasons to fear that an appreciable number do so under ground. The prevailing belief in the existence of sudden deaths is one of the chief causes of the terrible mistakes that lead to live burials. If this delusive idea were removed, those concerned, such as physicians, undertakers, relatives, and friends, would treat a person who unexpectedly took on the appearance of death as one needing careful attention by physician and nurse to bring him round to health again, as is usually done in cases of fainting. If trance were understood, doctors would be on the lookout for it; but, as it is not understood, it is called death, and we bury our mistakes under ground.

Dr. Hilton Fagge, while doubting whether there is any foundation for the strong fear which many persons entertain of being buried alive after supposed death, allows that there is danger in cases of sudden death. In his "Principles and Practice of Medicine," Dr. Fagge says: "The cases really requiring caution are some very few instances of persons found in the streets, or losing consciousness unexpectedly and in unusual circumstances."[1]

Dr. Léonce Lénormand, in "Des Inhumations Précipitées," p. 86, says that medical archives record details of a great number of apopletic cases revived after one, two, and three days' apparent death; and observes that the most celebrated physicians, both ancient and modern, agree in recommending delay in the burial of persons who succumb to this affliction.

---

[1] In the 3rd ed., by Dr. Pye Smith, the following occurs at p. 817 of vol. i., under "Trance" :—"These are the cases which have led to the popular belief that death is sometimes only apparent, and that there may be a danger of persons being buried alive; and it cannot be denied that a patient in such a condition might easily be allowed to die by careless or ignorant attendants, or might be buried before death."

Dr. Franz Hartmann, in his " Premature Burial," p. 11, quotes the following :—

"In the Bukovina, a young woman, in the vicinity of Radautz, died of spasms of the heart. They waited five days for the funeral, because no signs of putrefaction appeared. The clergyman then refused any longer delay, and the final arrangements for interment were made. Just as they were about to put the coffin into the grave, the sister of the deceased woman, who lived at another place, arrived, and begged to be permitted to see the dead body. Owing to her entreaties the coffin was opened, and as the woman saw the unaltered features of her sister, she asserted her belief that the supposed dead was still living. She procured a red-hot poker, and, in spite of the remonstrances of those present, she touched with it the soles of the feet of the corpse. There was a spasmodic jerk, and the woman recovered. The most remarkable thing was that the supposed dead woman had not been unconscious for a moment, but was able to describe afterwards all the details of what had taken place around her, from the moment when she was supposed to die up to the time of her recovery ; but she had looked upon all that like an unconscious spectator, and not experienced any sensation, nor was she able to give any sign of life."

In " Les Signes de la Mort," by Dr. E. Bouchut, p. 51, Dr. J. Schmid is cited for the case of a girl, seven years of age, who, while playing with her companions, fell suddenly down (as if struck by lightning), and died. There was paleness, absence of pulse, insensibility to all stimulus. Nevertheless, owing to the requests of the distressed parents, the apparently hopeless attempts at resuscitation were continued. After three quarters of an hour the girl gave a sigh and recovered.

The *Medical Record*, New York, 1883, vol. xxiii., p. 236, contains a paper on " Revivification " (in cases of sudden apparent death from heart-disease, and in the still-born), by S. Waterman, M.D., New York

Case 1, February, 1880.—Mr. B——, aged 84, suffered from valvular disease of the heart, and likewise from Bright's disease. " One morning, while I was sitting at his bedside and in friendly conversation with him, he being to all appearance in a very happy mood of mind, he suddenly fell back, his eyes became fixed and glassy, a deadly pallor crept over his countenance, respiration and the heart's action ceased simultaneously, and death seemed to have carried him off suddenly and unexpectedly. It was this suddenness of the event that impelled me to make efforts at revivification. Two nephews of Mrs. B——, who were fortunately in the house, were brought under requisition, and, under my direction, systematic artificial movements were carried on for nearly thirty minutes, when one deep inspiratory effort was made by the patient himself. Thus encouraged, we redoubled our efforts for ten minutes more ; other inspiratory efforts followed in quicker succession; the heart began to respond. Hardly audible at first, it acquired force and momentum ; it could now be felt at the wrist ; the deadly pallor passed away, the eyes lost their glassy, fixed aspect, sighs and groans could be heard, twitchings of the muscles of the arm and fingers could be distinctly felt, and the appearances of death made way for reanimated conditions. He lay unconscious for more than ten hours, respiration being hurried, and breathing stertorous, the heart's action wild and irregular. During the night he was delirious and restless ; toward morning all untoward symptoms subsided, and a quiet sleep followed the extreme restlessness. . . . He died six weeks afterwards, under symptoms of uræmic toxication. During these six weeks he had several other attacks—one very prolonged and almost fatal—in which artificial respiration was resorted to with the same success."

The editor of the *Manchester Criterion*, December 11, 1895, says:—" Many cases of sudden death have been entombed who were really alive, so far as the union of the body and soul is concerned. Sudden disappearance of life is very common, due to excessive weakness or a partial cessation of the heart's action ; and doctors should be very chary in giving death-certificates until it has been ascertained that decomposition has ensued. Many object to this delay, and on the approach of an indication of death, or apparent death, often hurry the body to the grave. We know of a young lady, for whom the shroud was bought, and the crape fastened on the door, who was restored to life."

SUDDEN DEATH.

Professor Alexander Wilder, M.D., in " Perils of Premature Burial," p. 16, says :—" In this country (America), however, the peril of interment before death has actually taken place is very great. For years past it has been a very common occurrence for persons in supposed good health to fall down suddenly, with every appearance of having died. We do not regard sudden death with terror, as it is so often painless, and exempts the individual from the anxiety and other unpleasant experiences which so often accompany a lingering dissolution. But there is a terrible liability of being prostrated by catalepsy, the counterpart of death, under such circumstances that those who have the body in charge will not hesitate about a prompt interment."

"The difficulty of distinguishing a person apparently dead from one who is *really* so has, in all countries where bodies are interred precipitately, rendered it necessary for the law to assist humanity. Of several regulations made on this subject, a few of the most recent may suffice—such as those of Arras in 1772; of Mantua in 1774; of the Grand Duke of Tuscany in 1775; of the Senechaussée of Sivrai in Poitou in 1777; and of the Parliament of Metz in the same year. . . . These edicts forbid the precipitate interment of persons who die suddenly. Magistrates of health are to be informed, that physicians may examine the body; that they may use every endeavour to recall life, if possible, or to discover the cause of death." — *Encyclopædia Britannica, quoted by John Snart in Apparent Death, 1824, pp. 81-82.*

CHAPTER XIII.

SIGNS OF DEATH.

THE absence of respiration is the most ordinary sign of
death, but at the same time perhaps the one most likely
to deceive. To ascertain whether breathing be entirely
suspended, it is a practice to hold a looking-glass to
the face.

> " Lend me a looking-glass ;
> If that her breath will mist or stain the stone,
> Why, then, she lives."—*King Lear*, Act v., Sc. 3.

The common belief is that, if the operations of the
heart or lungs be arrested for ever so brief a period,
they will never be resumed, and upon a hasty diagnosis
and perhaps a trifling experiment the person is declared
dead. It would appear presumptuous to attempt to
doubt or deny a theory so widely accepted by both the
lay and medical world, but numerous well-attested facts
show that the action of the vital organs, with life itself,
may occasionally be actually suspended, as proved by
the most rigorous tests known to science, and that
various forms of suspended animation taking on the
appearance of actual death are of not unfrequent occurrence. Scepticism, prejudice, and apathy on this subject
have led to thousands of persons being consigned to the
grave to return to consciousness in that hopeless and
dreadful prison.

One of the most distinguished physicians in London informed the author that, being called in to decide a case of apparent or real death, he had applied the stethoscope and failed to detect the faintest pulsation in the heart, and yet the woman recovered. The danger of premature burial he believed to be very real and by no means an imaginary one, and his opinions were well known in the profession.

THE RESPIRATORY TEST.

Sir Benjamin Ward Richardson, in his paper on " The Absolute Signs and Proofs of Death," in the *Asclepiad*, No. 21 (1889), vol. vi., p. 6, says :—

" About the existence of respiratory movements there is always some cause for doubt, even amongst skilled observers ; for so slight a movement of respiration is sufficient to carry on life, at what I have in another paper designated 'life at low tension,' the most practised eye is apt to be deceived."

" The cessation of the indications of respiratory function, although useful in a general sense, is not by any means reliable. It is quite certain that in poisoning by chloral, and in catalepsy, there may be life when no external movement of the chest is appreciable."—*Ibidem, pp. 13, 14.*

CARDIAC AND ARTERIAL FAILURE TEST.

" Equal doubt attends the absence of the arterial pulsations and heart sounds. It is quite certain that the pulses of the body, as well as the movements and sounds of the heart, may be undetectable at a time when

13

the body is not only not dead but actually recoverable."
—*Ibidem, p. 14.*

In a review of several works on the " Signs of Death "
in *The British and Foreign Medical and Chirurgical
Review*, vol. xv. [1855], p. 74, W. B. Kesteven writes
that Bouchut's test of the cessation of the action of
the heart for one or two minutes is not to be relied
upon as a certain sign of death. " M. Josat has
recorded several instances wherein newly-born children
have been most carefully examined during several
minutes without the detection of the slightest cardial
sound or movement, and yet these have rallied and
lived. M. Depaul has collected ten similar instances.
M. Brachet has recorded[1] an instance of a man in
whom neither sound nor movement of the heart could
be heard for eight minutes, and who, nevertheless,
survived. Another adult case is mentioned by Dr.
Josat as having been witnessed by M. Girbal, of
Montpellier. . . . Sir B. Brodie and others have
described children born without hearts. The circulation is maintained at one period of human life without
the aid of the heart. It is, besides, quite consistent
with the facts observed in hysterical and other conditions of the nervous system, that the action of the
heart, like that of other muscles, should be so extremely
feeble as not to be cognisable by any sound or impulse,
and yet it may have sufficient movement slowly to move
the blood through the system, whose every function and
endowment is suspended and all but annihilated. In
cases of catalepsy, and of authentic instances of apparent
death, the respiratory muscles have not been seen to

---

[1] *Bulletin Therap. Méd.*, tome xxvii., p. 371.

move, yet inspiration and expiration—however slowly and imperceptibly—must have taken place."

THE PUTREFACTIVE TEST.

Dr. Roger S. Chew, of Calcutta, whose personal experiences of apparent death are elsewhere recorded in this volume, says :—

"Numerous expedients have been suggested as means of ascertaining whether a body is really dead or whether the animation is temporarily suspended ; but, though these suggestions may collectively yield a correct diagnosis, still they are valueless when separately considered, and cannot compare with the 'putrefaction test.'"

In the "Principles and Practice of Medicine" of the late Dr. Hilton Fagge, edited by Dr. Pye-Smith, vol. i., p. 19, of the second edition, is the following :—

"In most cases there is no difficulty in determining the exact moment at which death occurs. But sometimes it cannot be fixed with certainty, and there are some altogether exceptional instances (though I have never myself met with one) in which for hours, or even for days, it remains uncertain whether life is extinct or merely suspended. *I believe that the only sign of death which is both certain to manifest itself in the course of a few days, and also absolutely conclusive and infallible, is the occurrence of putrefaction*, which is generally first indicated by discoloration of the surface of the abdomen. And in any case admitting of doubt, the coffin should not be closed until this has shown itself." (Italics ours.)

The *Medical Examiner*, Philadelphia, vol. vi., p. 610, says :—

"A recent French reviewer in the *Gazette Médicale*

closes a survey of the differences between real and
apparent death, by the following remarks :—'Experience,' says he, 'has shown the insufficiency of each of
these signs, with one exception — *putrefaction.* The
absence of respiration and circulation, the absence of
contractility and sensibility, general loss of heat, the
hippocratic face, the cold sweat spreading over the body,
cadaveric discoloration, relaxation of the sphincters, loss
of elasticity, the flattening of the soft parts on which
the body rests, the softness and flaccidity of the eyes,
the opacity of the fingers, cadaveric rigidity, the ex-
pulsion of alimentary substances from the mouth ;—all
these signs combined or isolated may present themselves
in an individual suffering only from apparent death.' "

Prof. D. Ferrier, in an article on "Signs of Death" in
Quain's " Dictionary of Medicine," pp. 327, 328, says :—

" It is not always easy to determine when the spark
of life has become finally extinguished. From fear of
being buried alive, which prevails more abroad than in
this country, some infallible criterion of death, capable
of being applied by unskilled persons, has been considered a desideratum, and valuable prizes have been
offered for such a discovery. The conditions most
resembling actual death are syncope, asphyxia, and
trance, particularly the last. We cannot, however, say
that any infallible criterion applicable by the vulgar has
been discovered."

The writer then proceeds to describe the various
symptoms usually considered to denote death. The
chief of these is putrefaction, but he observes that
putrefaction may occur locally during life, and general
septic changes may occur to some extent before death.

Dr. Gannal, in "Signes de la Mort," p. 31, says :—

"I share the opinion of the majority of authors who have written on this subject, and I consider *putrefaction* as the only certain sign of death." The author then shows that all other signs are uncertain, and adds "that it is possible, by taking certain measures, to wait until putrefaction is well manifest, without injuring the public health." If the attending medical practitioner could always be relied upon to look for any such combination of signs as above suggested, there would be much less danger of premature burial than at present almost everywhere prevails ; but personal investigation obliges the author deliberately to declare that these are looked for only in a comparatively few instances.

RIGOR MORTIS.

With reference to *rigor mortis*, one of the signs many physicians regard as infallible as putrefaction, and to which the *British Medical Journal* attaches much importance, I cite the following :—

Dr. Samuel Barker Pratt says that *rigor mortis*, which is regarded as an absolute proof of death, is in itself a life-action, caused by a gradual withdrawal of the nerve-forces from the body, and is distinctly akin to, and the same in effect as, the tightening of a muscle, and other similar physiological actions in the living body.

Dr. Roger S. Chew observes :—

"*Rigor mortis* is a condition that seldom or never supervenes in the hot weather in India, and is often a feature of catalepsy.

"Ecchymoses, or *post-mortem* stains, are sometimes of value, but very frequently they do not appear, even though there are strong evidences of putrefaction having set in, and in some cases this cadaveric lividity, as it is termed, may be the result of violence received before animation was suspended, and, the vital spark not having been extinguished though the body was apparently dead, echymosis had asserted itself as a process of life, and not death."

Ebenezer Milner, M.D.Edinb., L.R.C.S.E., observes in a paper on "Catalepsy or Trance" in the *Edinburgh Medical and Surgical Journal*, 1850, vol. lxxiv., p. 330:—

"Patients labouring under an intense and prolonged paroxysm of catalepsy have been supposed to be dead, and have been interred alive.

"There are numerous cases of this kind on record, and many more where individuals, after being laid in their coffins, have fortunately recovered from the attack before the period of interment. In such cases respiration is insensible, and the heart's action is almost in abeyance; the surface of the body is nearly cold, and presents the pallor of death; and the articulations are stiff. Although it is no doubt a difficult task to distinguish this state of trance from the state of death, yet a careful examination of the body, and time, would lead to a correct diagnosis. The limbs after death are first lax, then stiff, and ultimately lax again. The stiffness of the limbs, known as the cadaveric rigidity, or *rigor mortis*, lasts for a longer or shorter time, according to circumstances; the sooner it supervenes, the shorter is its duration, and conversely. Now the stiffness of the limbs accompanying this intense form of trance supervenes at once, and lasts as long as the paroxysm continues. This is consequently a valuable diagnostic sign."

It may be observed that only in rare and very exceptional cases is time allowed for careful and accurate diagnosis.

CADAVEROUS COUNTENANCE.

Anthony Fothergill, in "A New Inquiry," 1795, p. 92 :—

"Nor can even the cadaverous countenance be, separately considered, an infallible test of life's total extinction. Nay, even putrefaction itself, though allowed to be the most unequivocal sign of death, might chance to deceive us in that syncope which sometimes supervenes on the last stage of the confluent small-pox, sea-scurvy, or other highly putrid diseases."

REGARDING CLENCHED JAWS.

A. de Labordette, Chirurgien de l'Hôpital de Lisieux, states in a letter to the Secretary of the Royal National Lifeboat Institution :—

"I have collected manifold observations relating to persons drowned or asphyxiated, in whose case contraction of the jaws was remarked, and who were subsequently restored to life." Dr. Brown-Séquard concurred in this, and declared further that such contraction is rather a sign of life than of death. — *Lancet, 1870, vol. i., p. 436.*

THE DIAPHANOUS TEST,

for the discovery of which a prize was given by the French Academy of Medicine, is regarded by Sir B. Ward Richardson as of secondary importance. It has certainly failed in many instances.

The following communication on

THE PROPER VALUE OF THE DIAPHANOUS TEST OF DEATH,

by Edwin Haward, M.D.Edin., F.R.C.S.Eng., appears in the *Lancet* of June 10, 1893, p. 1404 :—

"A case has come lately under my observation in which the value of the diaphanous test of death has been illustrated at its just worth, and, as the matter is one of supreme practical moment, I think it may be considered deserving a brief notice in the pages of the *Lancet*. Readers of the *Lancet* need scarcely be informed that the diaphanous test consists in taking a hand of a supposed dead person, placing it before a strong artificial light, with the fingers extended and just touching each other, and then looking through the narrow spaces between the fingers to see if there be there a scarlet line of light. The theory is that, if there be such a line of scarlet colour, there is some circulation still in progress, and therefore evidence of vital action, whilst, if there be no illumination, then the circulation has ceased and death has occurred. The French Academy of Medicine was so impressed with the value of this test that it awarded, I believe, to the discoverer of it a considerable prize. The illustration I am about to give indicates, however, that this test must be received with the utmost caution. The facts run as follows :—I was called in January last to visit a lady seventy-three years of age, suffering from chronic bronchitis. She had often suffered at intervals from similar attacks during a period of twenty-five years. The present attack was very severe, and as she was obviously in a state of senile decrepitude her symptoms naturally gave rise to considerable anxiety. Nevertheless, she rallied and improved so much that after a few days my attendance was no longer required. I heard nothing more of this lady until February 6,—a period of three weeks,—when I was summoned early in the morning to see her immediately. The messenger told me that she had retired to bed in the usual way, and had apparently died in the night, but that she looked so life-like there was great doubt whether death had actually taken place. Within half an hour I was by her bedside; there was no sign of breathing, of pulse, or of heart-beat, and the hands, slightly flexed, were rather rigid, but the countenance looked like that of a living person, the eyes being open and life-like. I believed her to be dead, and that the rigidity of the upper limbs indicated commencing *rigor mortis ;* but this curious fact was related to me by a near relative, that once before she had passed into a death-like state, with similar symptoms, even to the rigidity of the arms and hands, from which state she had recovered, and after which she had always experienced the direst apprehension of being buried alive. Her anxiety, it will be easily conceived, was readily communicated to her relatives, who urged me to leave nothing undone for determining whether life was or was not extinct. Under the circumstances I suggested that Dr. (now Sir) Benjamin Ward Richardson, who has made the proofs of death a special study, should be summoned. He soon arrived, and submitted the body to all the tests in the following order :—1. Heart sounds and motion entirely absent, together with all pulse movement. 2. Respiratory sounds and movements

entirely absent.   3. Temperature of the body taken
from the mouth the same as that in the surrounding
air in the room, 62° F.   4. A bright needle plunged
into the body of the biceps muscle (Cloquet's needle
test) and left there shows on withdrawal no sign of
oxidation.   5. Intermittent shocks of electricity at different tensions passed by needles into various muscles
and groups of muscles gave no indication whatever of
irritability.   6. The fillet-test applied to the veins of the
arm (Richardson's test) causes no filling of veins on the
distal side of the fillet.   7. The opening of a vein to
ascertain whether the blood has undergone coagulation
shows that the blood was still fluid.   8. The subcutaneous injection of ammonia (Monteverdi's test) causes
the dirty brown stain indicative of dissolution.   9. On
making careful movements of the joints of the extremities, of the lower jaw, and of the occipito-frontals, *rigor
mortis* is found in several parts.   Thus of these nine
tests eight distinctly declared that death was absolute ;
the exception, the fluidity of the blood, being a phenomenon quite compatible with blood preternaturally fluid
and at a low temperature, even though death had
occurred.   10. There now remained the diaphanous test,
which we carried out by the aid of a powerful reflector
lamp, yielding an excellent and penetrating light.   To
our surprise the scarlet line of light between the fingers
was as distinct as it was in our own hands subjected to the
same experiment.   The mass of evidence was of course
distinctly to the effect that death was complete ; but, to
make assurance doubly sure, we had the temperature of
the room raised and the body carefully watched until
signs of decomposition had set in.   I made a visit

myself on a succeeding day to assure myself of this fact.

" The results of these experimental tests were satisfactory, as following and corroborating each other in eight out of the ten different lines of procedure ; but the point of my paper is to show the utter inadequacy of the diaphanous test, upon which some are inclined entirely to rely.   Sir Benjamin Richardson has reported an instance in which the test applied to the hand of a lady who had simply fainted gave no evidence of the red line ; she therefore, on that test alone, might have been declared dead.   In my case the reverse was presented ; the body was dead, whilst the red line supposed to indicate life was perfectly visible.   Hence the test might possibly lead to a double error, and ought never of itself to be relied upon.

" It is a question worthy of consideration whether the colouration observed was due to the fluid state of the blood after death ; it is not unreasonable to suppose so, but I prefer merely to offer the suggestion without further comment."

Dr. Gannal, in his " Signes de la Mort," p. 54, says :—

" The loss of transparency of the fingers is an uncertain sign, because with certain subjects it takes place some time before death ; next, because it does not always occur in the corpse ; and finally, because it exists under certain circumstances in sick persons—in intermittent fever, for example, when the skin loses colour, the hands get cold, and the nails blue, as happens at the onset of the fits."

Orfila, " Médicine Légale," vol. i., p. 478, 4th edit., observes :—

"This sign can be of no use, because it is easy to prove that the fingers of corpses placed between the eye and the flame of a candle are transparent, even when this experiment is made one or two days after death."

Sir Benjamin Ward Richardson read a paper before the Medical Society of London on "The Absolute Signs and Proofs of Death," published (in 1889) in No. 21 of the *Asclepiad.* The circumstance which originated his investigation was a case of the revival of an apparently dead child immediately before the funeral. Dr. Richardson has seen persons apparently dead, and presenting all the signs of death, but who were really living. Amongst these he cites the following :—

"A medical man found dead, as it was presumed, from an excessive dose of chloral. To all common observation this gentleman was dead. There was no sign of respiration ; it was very difficult for an ear so long trained as my own to detect the sounds of the heart; there was no pulse at the wrist, and the temperature of the body had fallen to 97° Fahr. In this condition the man had lain for some hours before my arrival ; and yet, under the simple acts of raising the warmth of the room to 84° Fahr. and injecting warm milk and water into the stomach, he rallied slowly out of the sleep, and made a perfect recovery."

More remarkable is the case of a man struck by lightning, details of which Sir Benjamin received, in 1869, from Dr. Jackson, of Somerby, Leicestershire.

"The patient reached his home in a state of extreme prostration, in which he lay for a time, and then sank into such complete catalepsy that he was pronounced to be dead, and heard the sound of his own passing bell from the neighbouring church; by a desperate attempt at movement of his thumbs he attracted the attention of the women engaged about him, and,being treated as one still alive, recovered, and lived for several years afterwards, retaining in his memory the facts, and relating them with the most consistent accuracy."

Medical practitioners tell us that the signs of death are quite easy and impossible to mistake. Dr. Richardson, who has had the best of reasons, as already shown, for observation and investigation, holds a different opinion, and enumerates the signs of death as follows :—

(1) Respiratory failure, including absence of visible movements of the chest, absence of the respiratory murmur, absence of evidence of transpiration of water vapour from the lungs by the breath.

(2) Cardiac failure, including absence of arterial pulsation, of cardiac motion, and of cardiac sounds.

(3) Absence of turgescence or filling of the veins on making pressure between them and the heart.

(4) Reduction of the temperature of the body below the natural standard.

(5) Rigor mortis and muscular collapse.

(6) Coagulation of the blood.

(7) Putrefactive decomposition.

(8) Absence of red colour in semi-transparent parts under the influence of a powerful stream of light.

(9) Absence of muscular contraction under the stimulus of galvanism, of heat, and of puncture.

(10) Absence of red blush of the skin after subcutaneous injection of ammonia (Monteverdi's test).

(11) Absence of signs of rust or oxidation of a bright steel blade, after plunging it deep into the tissues. (The needle test of Cloquet and Laborde.)

Sir Benjamin sums up as follows :—

"If all these signs point to death—if there be no indications of respiratory function ; if there be no signs of movement of the pulse or heart, and no sounds of the heart ; if the veins of the hand do not enlarge on the distal side of the fillet ; if the blood in the veins contains a coagulum ; if the galvanic stimulus fails to produce muscular contraction ; if the injection of ammonia causes a dirty brown blotch—the evidence may be considered conclusive that death is absolute. If these signs leave any doubt, or even if they leave no doubt, one further point of practice should be carried out. The body should be kept in a room, the temperature of which has been raised to a heat of 84° Fahr., with moisture diffused through the air ; and in this warm and moist atmosphere it should remain until distinct indications of putrefactive decomposition have set in."

Dr. Franz Hartmann, whose recent monograph[1] has excited much attention both in the English and American Press, observes :—

"Apparent death is a state that resembles real death so closely that even the most experienced persons believe such a person to be really dead. In many cases not even the most experienced physician, coroner, or undertaker can distinguish a case of apparent death

---

[1] "Premature Burial : An Examination into the Occult Causes of Apparent Death, Trance, and Catalepsy." By Franz Hartmann, M.D. Second Edition. London : Swan Sonnenschein & Co. (One Shilling).

from real death, neither by external examination nor by means of the stethoscope, nor by any of the various tests which have been proposed by this or that writer, for all those tests have been proved fallible, and it is now useless to discuss them at length, because many of the most experienced members of the medical profession have already agreed that there is no certain sign that a person is really and not apparently dead, except the beginning of a certain stage of putrefaction. All other tests ought to be set down as delusive and un-reliable."

In the Royal Decree issued by the Government for examining the dead in Würtemberg, dated January 24, 1882 (*Dienst-Vorschriften für Leichenschäuer*, Stuttgart, 1885), various signs and experiments for enabling the official inspector of deaths to ascertain if actual death has taken place are laid down. Among these are:—

(1) " The cessation of sensibility may be assumed if, on raising the eyelid, the pupil remains unaltered when a lighted candle is held close to it; or if pungent odours, such as those derived from onions, vinegar, sal-ammoniac, or severe friction of the chest, arms, or soles of the feet, the application of mustard, or burning tinder, or if sealing-wax dropped upon the chest produces no reaction, and particularly if in the latter case the skin does not blister.

(2) " The stoppage of the circulation of the blood, apart from the absence of heart beating, if, after tying a tight bandage around the arm, the veins do not swell up, upon the hands being firmly gripped; also if, upon pricking the lips, no blood escapes; furthermore, if, on holding the hand in front of a bright light (the diaphan-

ous test), the finger-tips are no longer translucent as in the living."

Nor should the inspector ever neglect to examine the heart to ascertain the complete absence of all sound, and to test the absence of breath by other experiments.

The rescript further adds that these experiments "may not furnish absolute proof of death," and describes what further proceedings to institute. These are referred to in this volume in the chapter devoted to Death Certification.

An editorial note in the *Lancet*, January 29, 1887, p. 233, shows the difficulty of distinguishing real from

APPARENT DEATH.

"It was only last year that we commented in our columns upon the 'signs of death,' drawing attention to the more important criteria by which a skilful observer may avoid mistaking cases of so-called suspended animation from actual disease. Quite recently two instances have been recorded, in which, if report be true, it would seem there is still room for maturing the judgment upon the question herein raised. At Saumur a young man afflicted with a contagious disease apparently died suddenly. His body was enshrouded and coffined, but as the undertaker's men were carrying the 'remains' to their last resting-place they heard what they believed to be a knocking against the coffin-lid, and the sound was repeated in the grave. Instead of testing at once the evidence of their senses, they, in accordance with judicial custom, sent for the Mayor, in whose presence the lid was removed from the coffin. Whereupon, to the horror of the spectators, it was observed that the dead man had only just succumbed to asphyxia. The above narrative seems on the face of it too ghastly to be true, especially as the occupant of the coffin must have been shut up in a space containing oxygen in quantity totally inadequate to sustain an approximation to ordinary breathing. But in cataleptic and similar states the organic functions are reduced to the lowest ebb, and history records several instances in which, for a time at least, the determination of the living state was a matter of uncertainty. In our issue of the 15th inst., p. 129, the reader will find an account of 'Post-mortem Irritability of Muscle,' in which the phenomenon was manifested in a marked degree two hours after death from a chronic wasting disorder— a condition which favours early extinction of vital action in muscle. It may be argued, then, with some show of reasonableness, that it is quite possible for the heart to stand still, as it were, and yet retain the power of action, although experience tells us but little on the question as regards the human subject. Experiments on the lower animals, however, show that over-distension of the right cavities of the heart causes cessation of cardiac contraction, and that relief from the distension may be followed by resumption of the function of contractibility. It must not be forgotten that an analogous condition is witnessed at times in patients suffering from capillary bronchitis or other physical states underlying acute distension of the right heart ; for, in these cases, venesection is not uncommonly instrumental in arresting the rapidly failing cardiac contractions. The second case of apparent death alluded to above happened in 'the land of big things.' An inhabitant of Mount Joy,

14

Paramatta, was believed to be dead, and his supposed remains were about to be committed to the earth, when a mourning relative startled the bystanders by exclaiming, 'I must see my father once more; something tells me he is not dead.' The coffin was taken from the grave to the sexton's tool-house, and there opened, and was found to contain a living inmate, who justified the presentiment of his son by 'slowly recovering.' As no mention is made in either case of the period that elapsed between the occurrence of apparent death and the body being placed in the coffin, or of the time during which the encasement lasted, special and minute criticism is uncalled for. Enough has been said on the subject to emphasize the exhortation, 'Get knowledge, and with all thy getting get understanding.'"

The *British Medical Journal*, of September 28, 1895, in a leading article on the " Signs of Death," says :—

" The question of the possibility of the interment of living beings has recently been exercising the minds of a portion of the public, whose fears have found expression in a series of letters to some of the daily papers. It is a matter of regret that so much irresponsible nonsense and such hysterical outpourings should find a place in the columns of our great daily press. No attempt at the production of evidence in support of their beliefs or fears has been made by the majority of writers, whilst the cases mentioned by the few are either the inventions of the credulous or ignorant, or are destitute of foundation. It cannot be said that the few medical men who have joined in this public correspondence have either contributed any useful information or have seriously attempted to allay the fears of the public.

One medical gentleman managed to earn for himself a cheap notoriety by employing, with very scanty acknowledgment of the source, copious extracts from Dr. Gowers' article on ' Trance ' in Quain's ' Dictionary of Medicine.'

" The possibility of apparent death being mistaken for real death can only be admitted when the decision of the reality of death is left to ignorant persons. We are quite unprepared to admit the possibility of such a mistake occurring in this country to a medical practitioner armed with the methods for the recognition of death that modern science has placed at his disposal. Moreover, even by the ignorant the reality of death can only be questioned during the period preceding putrefaction. During this period various signs of death appear which, taken collectively, allow of an absolute opinion as to the reality of death being given. To each of these, as a sign of death, exception may perhaps be individually taken, but a medical opinion is formed from a conjunction of these signs, and not from the presence of an individual one."

The writer must surely have overlooked the able treatises by Winslow, Kempner, Russell Fletcher, Hartmann, Gannal, and others, supported by evidence in the aggregate of thousands of cases of premature burial or narrow escapes, or have forgotten the dreadful cases which have appeared from time to time in the columns of the *British Medical Journal* itself. Commenting upon the case of a child nearly buried alive, this medical authority in its issue of October 31, 1885, under the head of " Death or Coma," sensibly refers to some of the difficulties in distinguishing apparent from real death as follows :—

"The close similarity which is occasionally seen to connect the appearance of death with that of exhaustion following disease, was lately illustrated in a somewhat striking manner. An infant, seized with convulsions, was supposed to have died about three weeks ago at Stamford Hill. After five days' interval, preparations were being made for its interment, when, at the grave's mouth, a cry was heard to come from the coffin. The lid was taken off, and the child was found to be alive, was taken home, and is recovering. Such is the published account of the latest recorded case of suspended animation. We need not now attempt a dissertation on the physical meaning of coma. It is well known that this condition may last for considerable periods, and may at times, *even to the practised eye*, wear very much the same aspect as death. In the present instance, its association with some degree of convulsion may easily have been mistaken by relatives, dreading the worst, for the rigid stillness of *rigor mortis*. This is the more likely, since the latter state is apt to be a transient one in infants, though it is said to be unusually well marked in death from convulsions. One cannot, however, help thinking that the presence of the various signs of death was not, in this case, very carefully inquired into. It is hardly possible that, had the other proofs as well as that of stiffening been sought for, they would have been missed. *It is true that hardly any one sign short of putrefaction can be relied upon as infallible.* In actual death, however, one may confidently reckon on the co-existence of more than one of these. After a period of five days, not one should have been wanting. Besides *rigor mortis*, the total absence of which, even in forms of

death which are said not to show it, we take leave to
doubt, the *post-mortem* lividity of dependent parts afford
sure proof, as its absence suggests a doubt, of death.
Then there is the eye, sunken, with glairy surface, flaccid
cornea, and dilated insensitive pupil.   Most practitioners,
probably, are accustomed to rely upon stethoscopic
evidence of heart-action or respiration.   These alone,
indeed, are almost always sufficient to decide the question of vitality, if they be watched for during one or
two minutes.   There is no information as to whether
the child so nearly buried alive was seen by a medical
man.   It is difficult to believe that, if it had been, some
sign of life would not have been observed.   Still, the
case is a teaching one, even for medical men, and warns
us to look for a combination of known tests where any
doubt exists as to the fact of death."   The italics are
ours.

Prof. Alex. Wilder, M.D., in "Perils of Premature
Burial," p. 20, says:—

"The signs of total extinction of life are not so unequivocal as many suppose.   Cessation of respiration
and circulation do not afford the entire evidence, for the
external senses are not sufficiently acute to enable us to
detect either respiration or circulation in the smallest
degree compatible with mere existence.   Loss of heat
is by no means conclusive ; for life may continue, and
recovery take place, when no perceptible vital warmth
exists."

M. B. Gaubert, in "Les Chambres Mortuaires
d'Attente," p. 187, Paris, 1895, says :—

"One of the most celebrated physicians of the Paris
hospitals, according to Dr. Lignières, declares that out

of twenty certified deaths, one only presented indubitable characteristics of absolute death."

The difficulty of diagnosis in many cases being allowed renders the obligation and necessity for a radical change in our methods of treating the supposed dead a very urgent one. Medical writers, whilst admitting the unsatisfactory nature of the current practice of medical certification, allege that the remedy lies with Parliament to make compulsory a personal medical inspection of the dead, and to allow a fee as compensation for the trouble. This, however, would be very far from meeting the difficulty. How many general practitioners would be willing to submit half-a-dozen, say, of the eleven tests of death formulated by Sir Benjamin Ward Richardson, in any given case, and if willing, how many, having regard to the fact that these tests are not taught in the Medical Schools, and form no part of the usual medical curriculum, would be competent to make them with the requisite skill? In most of the Continental States there are State-appointed surgeons to examine the dead, *médecins vérificateurs*, and in some of these—Würtemburg, for instance—the official is obliged to examine the corpse several times before his certificate is made out. But notwithstanding this careful official inspection, cases of premature burial and narrow escapes are telegraphed by *Reuter* and *Dalziel* every now and then to the English Press, as we have seen, and additional details, with the names and addresses of the victims, are furnished by responsible special correspondents.

The best proof that one can give of the uncertainty of the signs of death is the great divergence of opinion amongst medical experts. Dr. Gannal, in "Signes de la Mort," Paris, 1890, p. 27, observes :—" If any of these signs had presented characters of absolute certainty, it is unquestionable that the unanimity of authors would have recognised it ; now, there is none. One sign held to be good by some, is declared bad by others." Dr. Gannal affirms with iteration that there is only one unequivocal sign and proof of dissolution—decomposition. All authorities agree that whatever degree of doubt attends the ordinary appearances of death, none dispute that this amounts to a demonstration.

When standing round the bed of a sick patient, reduced to a state of coma or suspended animation, to which death is the expected termination, as soon as the doctor utters the fatal words "all is over," no one present thinks of doubting the verdict, or putting it to the test. Mr. Clarke Irvine, who has had a wide experience, writing in the *Banner of Light*, December 14, 1895, Boston, U.S., says :—

" I have known of hundreds of deaths in my experience, and never have I known of any instance wherein a bystander has doubted save once, and then the person supposed dead was revived, and is now living out in Colorado. The mere accident of a stranger coming in just previous to the enclosing in a coffin prevented the man from the awful fate of burial alive, so far as we can see.

" In one other, the supposed dead man came to life a little before the time set for his funeral, by the accident of some one seizing hold of his foot : he is still living, and a resident of this country. The case was widely published in the newspapers after he was interviewed by a reporter in Chicago, where the rescued man was visiting at the time of the great Fair. He is known as Judge William Poynter. I saw him a few days ago, and have heard him relate the experience.

"The case of the little girl who was rescued while the funeral was in progress, at St. Joseph, Missouri, I have already contributed to *The Banner*. These people were saved by a mere chance ; how many have passed underground forever, of whom nothing was ever suspected ! All through the country, people are dying or apparently dying, or falling into death-like trances daily, and being placed in their coffins *as a matter of course*, and hurried to and into their graves, *as of course* also— and in the very nature of things it must be and must have been that hundreds upon hundreds have been and are being consigned to that most awful of all the dooms possible. The horror of the thing is simply unspeakable."

OFFICIAL REGULATIONS FOR THE PREVENTION OF PREMATURE INTERMENT IN BAVARIA.

The following are extracts from the Police Regulations for the inspection of the dead, and the prevention of premature burial in Bavaria, and issued by the Royal State-Ministry for Home affairs :—

§ 4.

In public hospitals, penitentiaries, charitable or other similar homes or institutions, the duty of inspection falls upon the physician in chief.

Outside these institutions the inspectors must be chosen, in the first instance, from among physicians, after them surgeons, former assistants of military hospitals, and lastly, in default of such, from lay people. The latter must, however, be of undoubted respectability, and, before their appointment, must be properly instructed by the district physician, and subjected from time to time to an examination.

§ 6.

As a rule the inspection of dead bodies must be made once if by doctors, and twice if by laymen. In communities which possess a mortuary a *second inspection* has to be made, even though the regular inspection has previously been made by doctors or laymen.

§ 7.

The first inspection has to be made as soon as possible after death, and, where practicable, within twenty-four hours, and in cases described under § 6, sec. 2, at least before removal of the body to the mortuary.

The second inspection must take place just before burial.

§ 8.

The body, until the arrival of the Inspector, must be left in an undisturbed position, with the face uncovered, and free from closely-fitting garments.

The instructions of the Inspector, for the resuscitation of a body suspected of apparent death only, are to be followed most strictly.

§ 9.

The Inspector has to give a certificate of corpse inspection confirmatory of his inspection, but he must only issue the same if he has fully ascertained the actuality of death.

§ 10.

(1) As a rule the bodies must not be interred before the lapse of 48 hours, but not later than 72 hours, after death.

The Police Authorities may, however, at the recommendation of the Corpse Inspector, exceptionally grant permission for the burial before the expiration of 48 hours if a *post-mortem* dissection has taken place, also if decomposition has set in, and if on account of lack of room the body has to be preserved in an overcrowded habitation.

APPENDIX to the Police Instructions as to Corpse Inspection and time of Burial, of 20th November, 1885.

I.

The purpose of corpse inspection is to prohibit the concealment of deaths by violent means or resulting from medical malpractices ; to detect infectious diseases, and the establishment of correct death lists ; and particularly *to prevent the burial of people only apparently dead.* For this purpose each corpse is to be closely examined on the first inspection as to any signs of death, both in the front and the back of the body.

II.

The Inspectors have primarily to establish the actuality of death by observing and notifying all the symptoms accompanying or following the decease.

Indications of death may be noted :—

(1) If there is no indication of any pulsation noticeable, either in the region of the heart, at the neck, at the temples, or the forearm.

(2) If the eyelids when pulled asunder remain open, and the eyes themselves appear sunken into their sockets, dulled, and lustreless, also if the eyeballs feel soft and relaxed.

(3) If parts of the body are pale and cold, if chin and nose are pointed, if cheeks and temples are sunken.

(4) If the lower jaw hangs down and immediately drops again if pushed up, or if the muscles feel hard and stiff (rigidity).

(5) If the skin of the fingers held against one another, held towards light, do not appear reddish.

(6) If a feather or burning candle held against the mouth show no sign of motion, or if there is no sign of moisture upon a looking-glass held before the mouth.

(7) If on different parts of the body, particularly the neck, back, or posterior, or the undersurface of the extremities there are bluish-red spots (death spots) visible.

(8) If the skin, particularly at the sides of the stomach, show a dirty-green discoloration (decomposition spots).

The non-Medical Inspector has to observe at least all the symptoms 1 to 4.

In doubtful cases the Medical Inspectors are advised to test the muscles and nerves by electric currents.

IV.

If the inspection gives rise to suspicions of apparent death (Scheintod), the inspector must (if he is not himself a doctor) immediately call for the assistance of a practised physician, so as to establish the actual condition, and to adopt the necessary measures for resuscitation, as follows :—

(1) Opening of the windows, and warming the room.

(2) Efforts at artificial respiration.

(3) Applications of warm mustard-plaisters to the chest and the extremities.

(4) Rubbing with soft brushes, with cloths saturated in vinegar, or spirit of camphor, also with hot woollen cloths.

(5) Irritation of the throat with a feather.

(6) Smelling sal-ammoniac.

(7) Dropping from time to time a few drops of "extract of balm" or similar essences into the mouth.

Unless medical aid has meanwhile arrived, the application of these measures must be continued until the apparently dead comes back to life, and begins to swallow, in which case he ought to have warm broth, tea, or wine, or until there is absolutely no doubt as to the total ineffectiveness of all attempts at re-animation.

CHAPTER XIV.

DURATION OF DEATH-COUNTERFEITS.

THE differences observed in the length of time that persons have remained in this condition depended, doubtless, upon the constitutional peculiarities of the patients—such as strength or weakness—or upon the nature of the disease from which they may have suffered. Struve, in his Essay, pp. 34-98, says "that it depends upon the proportion of vital power in the individual. Hence children and young persons will endure longer than the aged. Also upon the nature of the element in which the accident happened, whether it contained greater or less proportion of oxygenated or carbonic acid gas, or other poisonous vapours. The latent vital power seems to be much longer preserved when animation has been suspended by cold. A man revived after being under snow forty hours. Persons apparently dead sometimes awake after an interval of seven days, as was the case with Lady Russell. . . . In the female sex, the suspension of vital power, spasms, fainting fits, etc., originating from a hysterical, feeble constitution, are not rare, nor is it improbable that the state of apparent death may be of longer duration with them ; nay, it may be looked upon as a periodical disorder, in which all susceptibility of irritation is extinguished." Struve further remarks, p. 98, "that the state in which the vital power is suspended, or in which there is a want of susceptibility of stimuli, consists of infinite modifications, from the momentary transient fainting fit, to a death-like torpor of a day's duration. The susceptibility of irritation may be completely suppressed, and the apparently dead may be insensible of the strongest stimuli, such as the operation of the knife, and the effects of a red-hot iron."

M. Josat, in " De la Mort et de ses Caractères," gives the result of his own observations in one hundred and sixty-two instances, in which apparent death lasted—

In 7	....	from 36 to 42 hours.
20	....	„ 20 to 36 „
47	....	„ 15 to 20 „
58	....	„ 8 to 15 „
30	....	„ 2 to 8 „

The order of frequency of diseases in which these occurred was as follows: — Asphyxia, hysteria, apoplexy, narcotism, concussion of the brain, the cases of concussion being the shortest.

The length of time a person may live in the grave will depend upon similar concomitant conditions; but all things considered, a person buried while in a state of trance, catalepsy, asphyxia, narcotism, nervous shock, etc., and in any of the other states that cause apparent death without passing through a course of disease, and that occur during his or her usual health, will have a longer struggle before life becomes extinct than one whose strength had been exhausted by an attack of sickness. Estimates of the duration of such a struggle differ considerably. Some writers believe that "however intense, it must be short-lived." As to the prolongation of the horrible suffering incident to such tragic occurrences, Dr. Léonce Lénormand, in his " Des Inhumations Précipitées," pp. 2-4, observes—" It is a mistake to think that a living person, enclosed in a narrow box, and covered with several feet of earth, would succumb to immediate asphyxiation."[1]

Dr. Charles Londe, in his " La Mort Apparent," remarks :—" It has been calculated that, after one quarter of the quantity of atmospheric air contained in the coffin—approximately estimated at one hundred and twenty litres—was exhausted, death would set in ;

---

[1] " Pour se convaincre de l'erreur où l'on tomberait en adoptant cette opinion populaire, il suffit de réfléchir d'abord qu'un cercueil n'est pas exactement moulé sur les proportions du corps qu'il contient ; que, par consequent, tous les intervalles sont remplis d'air respirable, en quantité très-grande, égale à-peu-près à un cube dont le côté aurait 50 centimètres de hauteur. Or, chaque inspiration absorbe environ 1,200 centimètres cubes d'air dont l'oxygène n'est employé dans l'hématose que pour sa cinquième partie, le reste étant rendu pendant l'expiration ; il en resulte donc que chaque inspiration ne consomme en réalité que 240 centimètres cubes. L'homme, à l'état normal, respire à-peu-près 800 fois par heure ; et, comme un cube de 50 centimètres de côté contient 125,000 centimètres cubes, on doit conclure que cette quantité d'air peut suffire à 520 inspirations normales, c'est à dire à soutenir la vie pendant près de trois quarts d'heure. Mais, d'un autre côté, il est démontré, en botanique, que l'air filtre dans la terre ; celui contenu dans le cercueil peut donc en partie se renouveler. On doit nécessairement tenir compte de la nature du terrain où le cercueil a été déposé ; s'il est sec, léger ou sablonneux, il laissera pénétrer, circuler pour, ainsi dire, l'air atmosphérique plus facilement, que des terres humides, grasses ou argileuses. Ajoutons enfin, que les quantités déterminées plus haut pourraient être réduites de plus de moitié, sans causer directement la mort. On voit donc qu'un homme peut vivre sous terre pendant plusieurs heures, et que ce temps sera d'autant plus court que le sujet sera plus pléthorique, c'est-à-dire prédisposé aux congestions cérébrales, puisque, dans ce cas, ses inspirations seront plus larges et plus frequentes."

therefore, it is quite certain that, if the shroud is thick, and the coffin well closed, and the grave impenetrable to the atmosphere, life could not last more than forty to sixty minutes after inhumation. But is not that a century of torture ? "

Some allowance should be made for the persistence of the vital energy, which continues after all atmospheric air is cut off. " Experiments on dogs show that the average duration of the respiratory movements after the animal has been deprived of air is four minutes five seconds. The duration of the heart's action is seven minutes eleven seconds. The average of the heart's action after the animal has ceased to make respiratory efforts is three minutes fifteen seconds. These experiments further showed that a dog may be deprived of air during three minutes fifty seconds, and afterwards recover without the application of artificial means." [1]

Prof. P. Brouardel, M.D., Paris, in " La Morte Subité," p. 35, observes that :—" A dog, placed in a common coffin, lived five to six hours ; but a dog occupies less room than a man, who, in such a coffin, when closed, would not have more than one hundred litres, so he would possibly live twenty minutes. I would not wish anybody to pass twenty such cruel minutes."

" Mr. Bernard, a skilful surgeon of Paris, certified that, in the parish of Riol, he himself, and several other bystanders, saw a monk of the Order of St. Francis, who had been buried for three or four days, taken from his grave breathing and alive, with his arms lacerated near

---

[1] Report on " Suspended Animation." By a Committee of the Royal Med. Chirur. Society, July 12, 1862.

the swathes employed to secure them ; but he died immediately after his releasement.   This gentleman also asserts that a faithful narrative of so memorable an accident was drawn up by public authority, and that the raising of the body was occasioned by a letter written from one of the monk's friends, in which it was affirmed that he was subject to paroxysms of catalepsy."—*The Uncertainty of the Signs of Death, by Surgeon M. Cooper. Dublin, 1748.*

In a volume, entitled " Information Relative to Persons who have been Buried Alive," by Heinrich Friedrich Köppen, Halle, 1799, dedicated to Frederick William III., King of Prussia, and Louise, Queen of Prussia, are the nine following amongst many other cases :—

" *England.*—Lady Russell, wife of a colonel in the army, was considered dead, and only through the tender affection of her husband was she saved from living burial. He would not allow her to be taken away until decomposition would absolutely force him to do so.   After seven days, however, in the evening, when the bells were ringing, the faithful husband had the triumph to see her eyes open and her return to full consciousness."

" *Halle, Germany.*— Medical Professor Junker, in Halle, a very humane man, had a corpse of a suicide—by hanging—delivered for dissection at his college.   He was placed on a table in the dissecting room, and covered with a cloth.   About midnight, while the professor was sitting at his writing-table in an adjoining room, he heard a great noise in the dissecting room, and fearing that cats were gnawing at the corpses, he went out, and saw the cloth in a disturbed condition, and on lifting it up found the corpse missing.   As all the doors and windows were closed, he searched the room, and found the missing one crouching in a corner, trembling with cold, in the terror of death.   He besought the professor for mercy, help, and means for escape, as he was a deserter from the army, and he would be severely punished if caught.   After consideration the kind professor clothed him, and

took him out of town at night as his own servant—passing the
guards—pretending to be on a professional visit, and set him free
in the country. Years afterwards he met the same man in Hamburg
as a prosperous merchant."

"*Leipsic.*—The wife of the publisher, Mathäus Hornisch, died,
and, according to the custom of the times, the coffin was opened
before being put into the ground. The grave-digger noticed golden
rings on her fingers, and in the following night went to the grave
to steal them—which he found was not easy to do—when suddenly she drew back her arm. The robber ran away frightened,
leaving his lantern at the grave. The woman recovered, but
could not make out where she was, and cried for help. No one
heard her ; so she got out of the grave, took the lantern, and went
to her home. Knocking at the door, the servant called to know
who it was. She replied, "Your mistress. Open the door ; I
am cold, and freezing to death." The master was called ; and
happily she was restored to her home again, where she lived for
several years longer."

"*Pavese, Italy*, 1787.—A clergyman was buried, and noises were
heard in his grave afterwards. Upon opening the grave and the
coffin, the man was found alive, and violently trembling with fright."

"*Paris*, 1787.—A carpenter was buried, noises were heard proceeding from his grave, and upon opening it he was found to be
breathing. He was taken to his home, where he recovered."

"*Stadamhof*, 1785.—A young, healthy girl, on the way to a
wedding, had an apoplectic stroke, as it was thought, and fell as if
dead. The following day she was buried. The grave-digger, who
was occupied near her grave that night, heard noises in it, and
being superstitious ran home in fright. The following morning
he returned to finish a grave he was digging, and heard the
whining again from the girl's grave. He called for help, the grave
was opened, when they found the girl turned over, her face
scratched and bloody, her fingers bitten, and her mouth full of
blood. She was dead, with evidences of most dreadful suffering."

"*France.*—Madame Lacour died after a long sickness, and was
buried in a vault of a church, with all her jewels on. Her maid
and the sexton opened the coffin the following night to steal the
jewellery, when some hot wax from the candle they were using

15

fell on the woman's face and woke her up. The robbers fled in fright, and the woman went back to her home. She lived many years afterwards, and had a son who became a priest, who in turn—inheriting his mother's nature—underwent a fate similar to her own."

"*Lyons, France.*—The wife of a merchant died. Two days after her seeming death, and just before the time set for her burial, her husband, who, it seems, had some doubts as to her death, had her taken from the coffin, and had a scarifier used in cupping applied in twenty-five places without bringing any blood, but the twenty-sixth application brought her to consciousness with a scream, and she recovered completely."

"*Cadillac.*—A woman had been buried in the morning. In the following morning whining was heard in her grave. It was opened, and the woman was found still alive, but she had mutilated half of her right arm and the whole hand. She was finally restored."

The *Spectator*, October 11, 1895, publishes particulars of a recent case of recovery, after three days' interment, in Ireland. See pp. 111, 112 in this volume.

Köppen's investigations led him to observe that— " Human life may appear to come to a stop, and no one can say it will not go on again, if time enough is allowed for it to do so. This even the most learned in medicine cannot explain away or deny ; and the greatest precaution should be taken before death is declared to exist."

CHAPTER XV.

THE TREATMENT OF THE DEAD.

THE following extracts from French, English, and American authorities, who have made the subject of premature burial one of patient research, show how the dead, or apparently dead, were treated in their respective countries at the time they wrote, and when no reforms had been instituted. Buffon, who wrote more than a century ago, said :—" Life often very nearly resembles death. Neither ten, nor twenty, nor twenty-four hours are sufficient to distinguish real from apparent death. There are instances of persons who have been alive in the grave at the end of the second, and even the third day. Why, then, suffer to be interred so soon those whose lives we ardently wished to prolong? Most savages pay more attention to deceased friends and relatives, and regard as the first duty what is but a ceremony with us. Savages respect their dead, clothe them, speak to them, recite their exploits, extol their virtues ; while we, who pique ourselves on our feelings, do not show common humanity ; we forsake and fly from our dead. We have neither courage to look upon or speak to them ; we avoid every place which can recall their memory."

In his " History of the Modes of Interment among Different Nations," pp. 191-193, Mr. G. A. Walker, surgeon, quotes the following observations, as deserving

consideration on the subject of premature interment :—
"On many occasions, in all places, too much precipitation
attends this last office ; or, if not precipitation, a neglect
of due precautions in regard to the body in general ;
indeed, the most improper treatment that can be
imagined is adopted, and many a person is made to
descend into the grave before he has sighed his last
breath.   Ancient and modern authors leave us no doubt
respecting the dangers or misconduct of such precipitation.   It must appear astonishing that the attention of
mankind has been, after all, so little aroused by an idea the
most terrible that can be conceived on this side eternity.
According to present usage, as soon as the semblance
of death appears, the chamber of the sick is deserted by
friends, relatives, and physicians ; and the apparently
dead, though frequently living, body is committed to
the management of an ignorant and unfeeling nurse,
whose care extends no further than laying the limbs
straight, and securing her accustomed perquisites.   The
bed-clothes are immediately removed, and the body is
exposed to the air.   This, when cold, must extinguish
any spark of life that may remain, and which, by a
different treatment, might have been kindled into flame ;
or it may only continue to repress it, and the unhappy
person afterwards revive amidst the horrors of the
tomb.

"The difference between the end of a weak life and
the commencement of death is so small, and the uncertainty of the signs of the latter is so well established,
that we can scarcely suppose undertakers capable of
distinguishing an apparent from a real death.   Animals
which sleep in the winter show no signs of life.   In

this case, circulation is only suspended ; but were it annihilated, the vital spark does not so easily lose its action as the fluids of the body, and the principle of life, which long survives the appearance of death, may re-animate a body in which the action of all the organs seems to be at an end. But how difficult it is to determine whether this principle may not be revived. . . . Coldness, heaviness of the body, a leaden, livid colour, with a yellowness in the visage, are all very uncertain signs. M. Zimmermann observed them all upon the body of a criminal, who fainted through dread of that punishment which he had merited. He was shaken, dragged about, and turned in the same manner as dead bodies are, without the least signs of resistance, and yet, at the end of twenty-four hours, he was recalled to life by means of the volatile alkali." Mr. Walker's history was written nearly sixty years ago, but the custom he deprecates still continues.

Dr. Moore Russell Fletcher, in his " Suspended Animation and Restoration," Boston, 1890, p. 19, speaking of the treatment of the dead in the United States, says :—" It is doubtful whether modern civilisation has much advanced the rites of burial, or the means of preventing interment before positive death. The practice now is, as soon as apparent death takes place, to begin at once preparing the body for burial ; the relatives and physician desert the room, pack it in ice or open the windows, thus banishing any possible chance of reviving or resuscitating any spark of vitality which may exist. No examination is ever made by the physician or the friends to see if there are even the faintest signs of life present. Under such circumstances, and with no attempts made at discovering whether any signs of life were still present (but a hasty burial instead), it is not strange that cases of premature interment frequently occur."

The Rev. Walter Whiter, in his "Dissertation on the Disorder of Death," 1819, p. 328, sensibly observes :— "The signs marked on the dying and the dead are fallacious. The dying man may be the sinking man, exhausted by his malady, or perhaps exhausting his malady, and fainting under the conflict. Exert all the arts which you possess, and which have been found not only able to resuscitate and restore the dying, but even the dead ; rouse him from this perilous condition, and suffer him not, by your supineness and neglect, to pass into a state of putrefactive death." And in p. 363 :— "If the humane societies had applied the same methods in various cases of natural death which they have adopted in the case of drowning, and if they had obtained a similar success in the cultivation of their art, the gloom of the bed of death would be brightened with cheering prospects, and would have become the bed of restoration and the scene of hope."

In this connection we may remark that no profession is more overcrowded at the present time than that of medicine, particularly in the United Kingdom, the English Colonies, and the United States. Hundreds of young men graduate from medical colleges every year, vainly seeking openings for a practice ; and some, for the purpose of gaining a livelihood, resort to expedients which the *Lancet* denounces as undignified,

unprofessional, and disgraceful.[1]   Then, again, the
number of nurses and of those qualifying for this
honourable vocation is already in excess of the
demand, and nursing institutions, under the keen competition to which they are subjected, are reducing their
charges.   Now, the care and treatment of the supposed
dead is an honourable vocation, offering a wide field for
the instructed physician and the tender and sympathetic
nurse, and if the appliances for resuscitation were always
at hand, as they should be, in every hospital, town-hall,
mortuary, police station, and in all large hotels and
churches, many lives now subjected to the risks of
premature burial would be saved.   While in London
there are two or more houses or retreats for the
dying, there is no place for the apparently dead but
a shunned and neglected coffin.   The time is not far
distant when the present mode of treating the dead
and the apparently dead—a practice born of superstition and fear, by which many are consigned to
premature graves — will be catalogued amongst the
barbarisms of the nineteenth century.

---

[1] The *British Medical Journal*, August 15, 1894, p. 381, reports a
"Discussion on the Overcrowding of the Profession," in which Dr.
Frederick H. Alderson says :—"The very crowded condition of the
medical profession concerns a very large body of the profession ; neither is
the evil limited to any particular section of it.   Our physicians are too
numerous, our surgeons alike too many, and our general practitioners are
legion."

CHAPTER XVI.

THOSE interested in the movement, if we are right in designating the widespread feeling of discontent by this name, are occasionally asked if the cases of premature burial are numerous, and what estimates, if any, have been made of them. We have no means of answering these queries. We do not even know the percentage of people who are subject to trance, catalepsy, shocks, stroke of lightning, syncope, exhausting lethargy, excessive opium-eating, or other diseases or conditions which produce the various death-counterfeits. Personal inquiries over a considerable portion of Europe, America, and the East prove that such cases are by no means of infrequent occurrence, and this is the deliberate conclusion of nearly all the authorities cited in this volume.

Dr. Chambers wrote in 1787—"Every age and country affords instances of surprising recoveries, after lying long for dead. From the number of those preserved by lucky accidents, we may conclude a far greater number might have been preserved by timely pains and skill."— *Cited in Mort Apparente et Mort Réelle, p. 17.*

In his introduction to the work above cited, "Information Relative to Persons who have been Buried Alive," by Henrich Friedrich Köppen, Halle, 1799, the author says:—"General Staff Medical Officer, D. O. in D., states that, in his opinion, one third of mankind are buried alive." This estimate is very obviously exaggerated, although many trustworthy experiences prove that a certain number of those who die have returned to consciousness in their graves. A great many are buried alive from ignorance of their relatives, who mistake coldness of the body, stoppage of the pulse and breathing, the colour of death, spots of discolouration, a certain odour, and stiffness of the limbs—which are only deceptive signs, not the signs of real death.

The very respectable Dr. Hufeland says:—"One cannot be too careful in deciding as to life or death, therefore I always advise a delay of the funeral as long as possible, so as to make all certain as to death. No wonder when those who are buried alive, and who undergo indescribable torture, condemn those who have been dearest to them in life. They will have to undergo slow suffocation, in furious despair, while scratching their flesh to pieces, biting their tongues, and smashing their heads against their narrow houses that confine them, and calling to their best friends, and cursing them as murderers. The dead should not be buried before the fourth day ; we even have examples that prove that eight days or a fortnight is too soon—as there have been revivals as late as that. I say every one should respect those who only seem to be dead. They should be treated gently, and kept in a warm bed for thirty-six hours."

Mr. John Snart, in his " Thesaurus," pp. 27, 28, London, 1817, says : — " The number of dreadful catastrophes, arising from premature interment, . . . that have been *discovered* only, or have transpired to man, *above ground*, both in ancient and modern times, conveys to every reflecting mind the fearful thought that they are but a *sample* (per synecdochen) out of such an incalculable host, perhaps one in a thousand."

Professor Froriép, quoted in Kempner's volume, says that—"In 1829, arrangements were made at the cemetery, New York, so as to bury the corpses in such manner as not to prevent them communicating with the outside world, in case any should have awakened to life; and among twelve hundred persons buried six came to life again." In Holland, the same author states, of a thousand cases investigated, five came to life before burial or at the grave. The Rev. J. G. Ouseley, in his pamphlet on "Earth to Earth Burial," London, 1895, estimates "that two thousand seven hundred persons at least, in England and Wales, are yearly consigned to a living death, the most horrible conceivable."

The Rev. Walter Whiter, in the "Disorder of Death," 1819, p. 362, calls attention to one of the reports (of Humane Societies) where the following passage occurs : "Monsieur Thieurey, Doctor Regent of the Faculty of Paris, is of opinion that one third, or perhaps half, of those who die in their beds are not actually dead when they are buried. He does not mean to say that so great a number would be restored to life. In the intermediate state, which reaches from the instant of apparent death to that of total extinction of life, the body is not insensible to the treatment it receives, though unable to give any signs of sensibility."

Maximilian Misson, in his "Voyage Through Italy," vol. i., letter 5, tells us "that the number of persons who have been interred as dead, when they were really alive, is very great, in comparison with those who have been, happily, rescued from their graves." He then proceeds to substantiate his statement by the recital of cases.

Dr. Léonce Lénormand, in his able treatise, "Des Inhumations Précipitées," has given his deliberate opinion that a thousandth part of the human race have been, and are, for want of knowledge, annually buried alive. This we regard as an under, rather than an over-estimate.

M. Le Guern, in his "Danger des Inhumations Précipitées," which has passed through several editions, declares that he has personally met with forty-six cases of premature burial in twelve years. He devoted thirty years to the study of the facts, and collected a list of two thousand three hundred and thirteen cases from various sources. He estimates the number of premature burials in France at two per thousand.

On February 27, 1866, the petition of M. Cornot was presented to the French Senate by M. de la Gueronnière, stating that a comparatively large number of persons are annually buried alive, which he supported by statistics. The author has tried to procure a copy of this petition, but these documents are not published by the State department.

The following appears in the *Lancet*, June 14, 1884, p. 1104 :—

"BURIED ALIVE.

"Sir,—That this is an incident that does happen, and frequently has happened, has for some years past been my firm conviction ; and during epidemics, particularly in the East, its possible contingency has frequently caused me much anxiety ; and when the burial has, for sanitary reasons, had to be very hurried, I always made it a rule to withhold my certificate unless I had personally inspected the body and assured myself of the fact of death.

"The reason and necessity for extreme caution in such matters were impressed vividly upon me some years ago, when visiting the crypt of the cathedral at Bordeaux, where two bodies were shown, to whom, I think it obvious, this most terrible of all occurrences must have happened; and I am unable to attribute the position in which they were found in their coffins, and the look of horror which their faces still displayed, to any action of *rigor mortis* or any other *post-mortem* change, but simply and solely to their having awakened to a full appreciation of their most awful position. In the case of one of these bodies, which was found lying on its side, the legs were drawn up nearly to a level with the abdomen, and the arms were in such a position as to convey the impression that both they and the legs had been used in a desperate, but futile, attempt to push out the side of the coffin; whilst the look of horror remaining on the face was simply indescribable. In the other case, the body was found lying on its face, the arms extended above the head, as if attempting to push out the top of the coffin. In the year 1870 these two bodies were still on view; and the attendants used to dwell at some length upon the horrors of being interred alive. It appears that some years prior to 1870, in making excavations in a church-yard in the immediate vicinity of the cathedral, the workmen came upon a belt of ground that apparently was impregnated with some antiseptic material, as all the bodies within this belt, to the number of about two hundred, were found to be almost as perfect as when they were buried ; of these a selection appears to have been made ; and at the time I mention about thirty or forty were exhibited, propped up on iron frames, in the crypt of the cathedral. The impression left on my mind at the time was that, if out of two hundred bodies so discovered there could be two in which, to say the least, there is a strong probability of live interment, this awful possibility was a thing that should receive more attention than is generally devoted to it.—I am, Sir, your obedient servant,

"H. S.

" Bayswater, June 10, 1884."

Protests against the present state of the law in France are very frequent.   M. Gaubert in " Les Chambres Mortuaires d'Attente," page 80, says : " During the monarchy of July petitions have not ceased to come in from all parts of France to the Chamber of Deputies." For a great number of years, said the Deputy Varin, in the sitting of April 10, 1847, every year petitions having the same object (the prevention of premature burial) are presented to the Chambers and referred to the Ministry. What has been done, however ? Nothing ! Again M. Gaubert in p. 88, referring to resolutions of the General Councils of the Departments, observes : " That under the movement of protest, which we are examining and find particularly serious, is shown the widespread character which it assumes.   It is, indeed, from all parts of France, and under every form, that the sad com- plaints of the public (for the prevention of premature burial) arrive at the office of the Minister of the Interior. Those protests adopted by the General Councils (of

Departments) were not the less numerous nor the less conspicuous in important places. Many of those who take the trouble to petition or draw up resolutions have been prompted to action by melancholy experience of such catastrophes in their own families."

M. Gaubert in "Les Chambres Mortuaires d'Attente" (Paris, 1895), pp. 193-195, says that in France there are in round numbers thirty-six thousand Communes, and it is beyond doubt that in every one of these will be found cases of premature burial. Communes with a population of eight hundred have even several. Dr. Pineau has recorded twelve in the single Commune of Fontenay-le-Comte in Poitou. In the large towns, especially in those which have great hospitals, the proportion is more considerable. In Paris, Dr. Rousseau, verificateur of the dead, in 1853 wrote: "Le médecin n'est jamais appelé que pour constater la mort apparente." M. Gaubert declares that he would not be far from the truth in estimating the number of victims to apparent death at eight thousand a year, and asks if France be so rich in population as to be able to pay such an enormous tribute. Dr. Josat, lauréat de l'Institut, declares that a considerable number of people refuse to visit France through fear that they might be overtaken by apparent death and precipitately buried alive.

The *Undertakers' Journal*, July 22, 1889, the editor of which has exceptional opportunities of knowing the true facts, observes: "It has been proved beyond all contradiction that there are more burials alive than is generally supposed. Stories of these cases are numerous. Five cases are reported on p. 85 of this same issue, one the wife of a well-known tradesman at St.

Leonards, medically pronounced dead, but who revived before it was too late. Many undertakers could describe similar experiences."

Dr. Roger S. Chew, of Calcutta, in reply to the author's inquiries while in India in the early part of the year (1896) says : " There are hundreds of instances on record where from some cause, as syncope, shock, chloroform, hysteria, or other condition not clearly understood, the powers of life assumed a static condition in which oxidation was completely arrested, carbonification was held in abeyance, and nitrification maintained at positive rest, with the consequence that the vital functions have passed into a condition of hibernation or apparent death so closely simulating real or absolute death as to render differential diagnosis an almost impossibility, and to lead to the interment or cremation while yet alive of a body apparently dead."

Dr. Franz Hartmann, of Hallein, Austria, whose book, " Buried Alive," is now being translated into French, has collected seven hundred cases of premature burial and narrow escapes, several of which have occurred in his own neighbourhood, and is of opinion that the actual danger to every member of the human family is of serious proportions, and that the subject should not be trifled with. He is a strong advocate for cremation as offering the easiest practical method of prevention.

It will have been noticed that whenever the subject of premature burial has been introduced in an influential journal published in England, the United States, or the Continent, one contribution follows another in quick succession, by persons furnishing particulars of cases of trance, catalepsy, and of narrow escapes from living burial. The Paris *Figaro* opened its columns two years ago for this subject, and in fifteen days received four hundred letters from all parts of France. When we consider that nearly all the reported cases of resuscitation have come about spontaneously and independently of human intervention, it becomes evident, owing to our ignorance and apathy, that cases of premature burial are far from infrequent, and our church - yards and cemeteries, like those examined by Dr. Thouret in Paris, are probably the silent witnesses of unnumbered unspeakable tragedies. Immediate legislation is called for to remedy a national evil, and to remove the feeling of disquietude which extensively prevails.

CHAPTER XVII.

EMBALMING AND DISSECTIONS.

AN intelligent and observing correspondent writes to the author that "under the prevailing custom of embalming in vogue in the United States, it is almost impossible to have a living burial, as the injection of the fluids used in the operation would prevent revival and make death certain. Of course, the class denominated 'poor folks,' who cannot afford this security, have to take their chances with the mysteries of trance and other forms of apparent death, as well as with ignorance, indifference, and unseemly haste, that seem to encompass a man at a time when he is in need of the most considerate care."

Embalming is no doubt preferable, as was thought by the late Lady Burton, to the risks, prevailing in almost all countries, of burial before careful medical examination, for the reason that it is better to be killed outright by the embalmer's poisonous injections, or even to come to life under the scalpel of the anatomist, than to recover underground. A leading New York investigator has openly declared his belief that a considerable number of human beings (supposed by their relatives to be dead, but who are really only in a state of death trance) are annually killed in America by the embalming process.

16

EMBALMING.

In the second edition of Dr. Curry's "Observations on Apparent Death," 1815, p. 105, the case is cited of William, Earl of Pembroke, who died April 10, 1630. When the body was opened in order to be embalmed, he was observed, immediately after the incision was made, to lift up his hand.

F. Kempner, in "Denkschrift," p. 6, says :—

"Owing to some great mental exitement, the Cardinal Spinosa fell into a state of apparent death. He was declared to be dead by his physicians, and they proceeded to open his chest for the purpose of embalming his body. When the lungs were laid open, the heart began to beat again ; the cardinal returned to consciousness, and was just able to grasp the knife of the surgeon when he fell back and died in reality." [1]

[1] Quoted by Dr. Franz Hartmann in "Premature Burial."

The *Journal de Rouen*, Aug. 5th, 1837, relates the following :—

"Cardinal Somaglia was seized with a severe illness, from extreme grief ; he fell into a state of syncope, which lasted so long that the persons around him thought him dead. Preparations were instantly made to embalm his body, before the putrefactive process should commence, in order that he might be placed in a leaden coffin, in the family vault. The operator had scarcely penetrated into his chest when the heart was seen to beat. The unfortunate patient, who was returning to his senses at that moment, had still sufficient strength to push away the knife of the surgeon, but too late, for the lung had been mortally wounded, and the patient died in a most lamentable manner."

Dr. Hartmann in "Premature Burial," p. 80, says :—

"The celebrated actress Mlle. Rachel died at Paris, on 4th

January, 1858. After the process of embalming her body had already begun, she awoke from her trance, but died ten hours afterwards owing to the injuries that had been inflicted upon her."

The *Celestial City*, New York, June 15, 1889, records:—

"MRS. BISHOP'S EXPERIENCE.

"Mrs. Eleanor Fletcher Bishop, the mother of the celebrated mind-reader, has a thrilling experience of her own regarding the horrors of being railroaded into the grave. Anent the unseemly haste exercised by the doctors who made the autopsy on her son, the old lady stated what terrible perils she at one time barely escaped. ' I am subject to the same cataleptic trances in which my boy often fell,' said Mrs. Bishop. 'One can see and hear everthing, but speech and movement are paralyzed. It is horrible. For six days, some years ago, I was in a trance, and saw arrangements being made for my funeral. Only my brother's determined resistance prevented them from embalming me, and I lay there and heard it all. On the seventh day I came to myself, but the agony I endured left its mark forever.'"

Dr. P. J. Gibbons, M.A., says :—

"In my mind there is no doubt that bodies in which life is not extinct are embalmed. To prevent the embalming of live bodies in cases where doubt exists, my method for resuscitation should be resorted to. If success does not follow, death has taken place. When one in whom the vital spark may possibly not yet have fled is found, two objects should be aimed at, viz., first, to restore breathing, and, second, to promote warmth and circulation."—*The Casket*, Rochester, New York, April 1, 1895.

The Select Committee of the House of Commons appointed in 1893 to enquire into the subject of Death

Certification, suggests in their report that in all cases where it is desired to embalm a dead body an authorisation should be obtained from the Home Secretary. This is probably intended to prevent concealing cases of death by poisoning. The Select Committee might very well have extended its recommendations to the need of verifying the death before the embalmer was allowed to exercise his art on the subject. Legislation in the United States, where embalming is extensively practised among well-to-do people, is a matter of urgent necessity. The author is aware of only one town where the city ordinance enforces such verification before permitting burial.

Mr. M. Cooper, surgeon, in his admirable little volume "The Uncertainty of the Signs of Death," London, 1746, p. 196, observes that "those who are dissected run no risk of being interred alive. The operation is an infallible means to secure them from so terrible a fate. This is one advantage which persons dissected have over those who are, without any further ceremony, shut up in their coffins."

The following from Ogston's *Medical Jurisprudence*, p. 370, is a case in point (quoted by the *Lancet*) :—" In October, 1840, a servant girl, who had retired to bed apparently in perfect health, was found the following morning, as it appeared, dead. A surgeon who was called pronounced her to have been dead for some hours. A coroner's inquest was summoned for four o'clock, and the reporter and the surgeon who had been called in to the girl were ordered to inspect the body previous to its sitting. On proceeding to the house for this purpose at two o'clock, the inspectors found the girl lying in bed in an easy posture, her face pallid, but placid and composed, as if she were in a deep sleep, while the heat of the body had not diminished. A vein was opened by them, and various stimuli applied, but without affording any sign of resuscitation. After two hours of hesitation and delay, a message being brought that the jury were waiting for their evidence, they were forced to proceed to the inspection. In moving the body for this purpose, the warmth and pliancy of the limbs were such as to give the examiners the idea that they had to deal with a living subject! The internal cavities, as they proceeded, were found so warm that a very copious steam issued from them on exposure. All the viscera were in a healthy state, and nothing was detected which could throw the smallest light on the cause of this person's death." Tidy (*Legal Medicine*), part i., p. 140, remarks thereon—"A mistake had no doubt been made in this case, as its warmth was not caused by decomposition."

In the *Cyclopædia of Practical Medicine*, edited by Sir John Forbes, M.D., and others, 1847, vol. i., pp. 548-9, we find the following :—" Nothing is more certain than death ; nothing is more uncertain at times than its reality ; and numerous instances are recorded of persons prematurely buried, or actually at the verge of the grave before it was discovered that life still remained ; and even of some who were resuscitated by the knife of the anatomist. . . . Bruhier, a celebrated French physician, who wrote on the uncertainties of the signs of death in 1742, relates an instance of a young woman upon whose supposed corpse an anatomical examination was about to be made when the first stroke of the scalpel revealed the truth ; she recovered, and lived many years afterwards. The case related by Philippe Pue is somewhat similar. He proceeded to perform the Cæsarean section upon a woman who had to all appearance died undelivered, when the first incision betrayed the awful fallacy under which he acted. . . . 'There is scarcely a dissecting-room that has not some traditional story handed down of subjects restored to life after being deposited within its walls. Many of these are mere inventions to catch the ever greedy ear of curiosity; but some of them are, we fear, too well founded to admit of much doubt. To this class belongs the circumstance related by Louis, the celebrated French writer on medical jurisprudence. A patient who was supposed to have died in the Hospital Salpétrière was removed to his dissecting-room. Next morning Louis was informed that moans had been heard in the theatre ; and on proceeding thither he found to his horror that the supposed corpse had revived during the night, and had actually died in the struggle to disengage himself from the winding sheet in which he was enveloped. This was evident from the distorted attitude in which the body was found. Allowing for much of the fiction with which such a subject must ever be mixed, there is still sufficient evidence to warrant a diligent examination of the means of discriminating between real and apparent death ; indeed, the horror with which we contemplate a mistake of the living for the dead should excite us to the pursuit of knowledge by which an event so repugnant to our feelings may be avoided. . . . If life depends upon the presence of a force or power continually opposed to the action of physical and chemical laws, real death will be the loss of this force, and the abandonment of organised bodies to these agents; while apparent death will be only the suspension of the exercise of life, caused by some derangement of the functions which serve as instruments of vital action. This suspension must have been lost for a considerable time, if we may judge by the cases collected by credible authors, to some of which we have alluded, and by the numerous instances of drowned persons restored to life after long submersion. From this definition of life and death, it would follow that putrefaction is the only evidence of real death.' . . . The absence of the circulation of the blood has been looked upon as a certain indication of death ; but this test is not much to be depended on, for it is well known that persons may live even for hours in whom no trace of the action of the heart and arteries can be perceived."

Le Guern, in "Du Danger Des Inhumations Préci-pitées," chap. iv., p. 24, relates that "The Abbé Prévost was found in the forest of Chantilly perfectly insensible. They thought him dead. A surgeon proceeded to make a *post-mortem ;* but hardly had he put the scalpel in the body of the unfortunate victim before the sup-posed corpse uttered a cry, and the surgeon realised the mistake he had made. Prévost only became conscious to feel aware of the horror of the death by which he perished."

Dr. Franz Hartmann, in his "Premature Burial," p. 80, has the following :—.

"In May, 1864, a man died very suddenly at a hospital in the State of New York, and, as the doctors could not explain the cause of death, they resolved upon a *post-mortem* examination, but, when they made the first cut with the knife, the supposed dead man jumped up and grasped the doctor's throat. The doctor was terrified and died of apoplexy on the spot, but the "dead" man recovered fully.

Brigade-Surgeon W. Curran in his 8th paper, entitled "Buried Alive," relates the following:—"At the Medical College at Calcutta, on the 1st of February, 1861," so writes my friend as above, "the body of a Hindu male, about 25 years of age, was brought from the police hospital for dissection. . . . It was brought to the dissecting-room about 6 a.m., and the arteries were injected with arsenical solution about 7. At 11 the prosector opened the thorax and abdomen for the purpose of dissecting the sympathetic nerve. At noon Mr. Macnamara distinctly saw the heart beating; there was a regular rythmical vermicular action of the right auricle and ventricle. The pericardium was open, the heart being freely exposed, and lying to the left in its natural position. The heart's action, although regular, was very weak and slow. The left auricle was also in action, but the left ventricle was contracted and rigid, and apparently motionless. These spontaneous contractions continued till about 12.45 p.m., and, further, the right side of this organ contracted on the application of a stimulus, such as the point of a scalpel, &c., for a quarter of an hour longer."—*Health*, May 21st, 1886, p. 121.

Bruhier in his work, "Dissertation sur l'Incertitude de la Mort et l'Abus des Enterrements," records a number of cases of the supposed dead who, after burial, were revived at the dissecting table, together with fifty-three that awoke in their coffins before being buried, fifty-two persons actually buried alive, and seventy-two other cases of apparent death. This was at a time when body-snatching was in vogue, and it is a curious comment on our civilisation to be compelled to admit that a subject of trance or catalepsy during the last or the early part of the present century had a better chance of escape from so terrible a fate than now, when the vocation of the resurrection-man has become obsolete.

CHAPTER XVIII.

DEATH-CERTIFICATION.

A SELECT COMMITTEE of the House of Commons, under the chairmanship of Sir Walter Foster, M.D., was appointed on March 27, 1893, to inquire into the subject of death-certification in the United Kingdom. Fourteen sittings were held, and thirty-two witnesses examined. All the witnesses practically agreed as to the serious defects in the law, and a number of recommendations were made. It was shown that in about four per cent. of the cases the cause of death was ill-defined and unspecified, many practitioners having forms specially printed for their own use, in which all mention of medical attendance is omitted, the object being to enable the doctor to give certificates in cases which he has never attended. Numerous deaths attended by unqualified practitioners were certified by qualified practitioners who had probably never seen the cases; and deaths were certified by medical practitioners who had not seen the patient for weeks or months prior to death, and who knew only by hearsay of the deaths having occurred. Deaths were also certified in which the true cause was suppressed in deference to the feelings of survivors; these last in particular are reported to be very numerous.

In Q. 2552-83, remarkable evidence was produced as to the reckless mode of death-certification. One medical witness testified that he saw a certificate of death, signed by a registered medical practitioner, giving both the fact and the cause of death of a man who was actually alive at the time, and who lived four days afterwards, with facts of even a more startling character described as " murder made easy." It was pointed out that fraud and irregularity in giving false declarations of death are by no means infrequent. Various other matters are treated, and the following are some of their recommendations :—

1. That in no case should a death be registered without the production of a certificate of the cause of death by a registered medical practitioner, or by a coroner after inquest, or, in Scotland, by a procurator-fiscal.

2. That in each sanitary district a registered medical practitioner should be appointed as public medical certifier of the cause of death in cases in which a certificate from a medical practitioner in attendance was not forthcoming.

3. That a medical practitioner in attendance should be required, before giving a certificate of death, to personally inspect the body, but if, on the ground of distance, or for other sufficient reason, he is unable to make this inspection himself, he should obtain and attach to the certificate of the cause of death a certificate signed by two persons, neighbours, verifying the fact of death.

4. That medical practitioners be required to send certificates of death direct to the registrar instead of handing them to the relatives of the deceased.

5. That a form of certificate of death should be prescribed, and that medical practitioners should be required to use such form.

From the *Times*, May 23, 1896 :—

DEATH-CERTIFICATION.

At the special meeting of the Metropolitan Counties Branch of the British Medical Association, held last night at the Museum of Practical Geology, Jermyn Street, the subject of an improvement

in the present procedure in death-certification and registration
came up for discussion.   Sir W. Priestley, M.P., president, took
the chair.

Sir Henry Thompson moved the following resolution :—"Considering that a Select Committee of the House of Commons has
in 1893 made an extended inquiry into the subject of death-
certification and registration on the plan now followed in this
country, and has reported that it manifestly fails to accomplish the
purpose for which it was designed, this meeting is of opinion that
Her Majesty's Government should be respectfully memorialised to
bring in a bill as soon as possible to give effect to an improved
procedure in general accordance with the suggestions offered in
the Committee's report."   He said that, during the last twenty
years or more. circumstances had not unfrequently occurred to
attract public attention to the existence of grave defects in the
system of death-certification adopted in this country, whether
regarded as a safeguard against criminal attempts on life, or as a
means of forming trustworthy records of disease for scientific purposes.   From the Registrar-General's report for England and Wales
for the year 1892, it was shown that in fifteen thousand cases of death
no inquiry had been made as to its cause. and that no certificate had
been obtained from any source—a number amounting to nearly three
per cent. on the total returned for the year.   On the same authority
it appeared that in twenty-five thousand more, or four and a half
per cent., the cases "were so inadequately certified as not to be
classifiable," making together a class of seven and a half per cent. in
which no evidence of any value as to the cause of death existed.
After what had already been done in the matter, all that appeared
to be necessary at present seemed to him to be that they should
forward a memorial to the Home Secretary, with a request that he
would consider the important work which had been already done
by the Select Committee, and, if he saw fit, take steps to embody
their recommendations in an Act of Parliament, for the purpose of
giving the country a greatly improved procedure in exchange for
that at present employed.   Dr. Isambard Owen, in the absence of
Dr. Farquharson, M.P., seconded the resolution, and asserted that
the State now winked at an exceedingly loose system of death-
certification, since under the present procedure it was possible for

a medical man to give a death-certificate on a patient whom he might not have seen for an interval of several weeks, and perhaps months.   The resolution was supported by Dr. Nelson Hardy, Dr. Alderson, Dr. Hugh Woods, Dr. Sykes, and others, and was unanimously adopted.

A well-known physician in large practice, writing to the author from a Midland town, October 10, 1895, says :—"Medical men, attending patients seriously ill, accept the statement of the friends that the patient died in the night, and give a certificate at once, without any inspection of the body.   This is the regular practice."

In Ireland matters are no better, and clergymen and others, with whom the author has been in correspondence, say they are much worse, and the danger of premature burial is, if possible, greater than it is in England.   The Rev. W. Walters, writing from Ventry Parsonage, Dingle, Ireland, September 16, 1895, says :— "In Ireland interment usually takes place the day after decease, and no certificate as to the cause of death is ever required.   There is no safeguard whatever, and amongst the ignorant poor I fear premature burial is terribly frequent."

A prominent medical officer of health, having charge of a populous metropolitan parish, wrote to the author, October 8, 1895, in reply to inquiries :—"When a doctor attends a patient in an illness, and the patient dies, he usually accepts the word of the friends as to the facts of death, and if they are poor, or in moderate circumstances, he grants the certificate in the ordinary way. If he is satisfied as to the cause of death he dare not refuse the certificate.   You will see by the form I send you that *he need not actually satisfy himself that the patient is dead;* if he is not satisfied he writes, ' As I am informed,' in the space left for the words. . . . On one occasion I was directed by a lady to drive a very long hat-pin through her heart after death, to ensure that she should not be buried alive. I have given so little attention to the matter that I cannot say if the Continental practice in this respect is better than ours. *Signs of decomposition are, I believe, the only ones of any real value.* The form of certificate of death referred to is marked, ' Printed by authority of the Registrar-General,' and a request marked ' N.B.' is to read the suggestions on page ii. In this other form, which is entitled ' Suggestions to medical practitioners respecting certificates of the cause of death,' elaborate instructions are set forth under ten separate clauses, with examples showing in what way the death-certificates are to be filled up, but not one word of instruction or caution as to the fact of death—whether it be real or apparent—the absolute signs of death, or the steps to be taken in doubtful cases, or in the various forms of suspended animation, such as coma, trance, catalepsy, etc."

The *Times,* January 19, 1878, p. 9, foot of column 6, reports a singular case in point :—

" PREMATURE.—A poor woman lay very ill in her scantily-furnished home in Sheffield. The doctor was sent for, and came. He at once saw that hers was a very grave case, and that she had, as he thought, little chance of recovery, even if she could get the nourishment her illness required. As he was about to leave, the question was put, ' When should we send for you again, doctor?' 'Well,' was the reply, as he looked at the poor woman and then at her wretched surroundings, ' I

don't think you need send for me again.   She cannot
possibly get better ; and to save you further trouble I'll
just write you out a certificate for her burial.'   And he
did.   After the doctor departed the woman—women
always were wilful—got better rapidly.   She has now
completely recovered, and goes about carrying her burial
certificate with her.—*Sheffield Telegraph.*"

Dr. Charles Cameron, M.P., in moving the introduction of the Disposal of the Dead (Regulation) Bill, in
the House of Commons, on April 30, 1884, said :—" A
very large number of our population die without any
medical attendance at all, or at least without having
ever received sufficient medical attention to enable a
certificate of the cause of death to be given worth the
paper on which it is written.   In many of these cases
some sort of worthless certificate is procured and presented to the registrar, but many thousands of persons
are each year buried in the United Kingdom without
even this formality."

The contrast between the laxity at home and the
regulations laid down by authority in Würtemburg,
Bavaria, and other Continental States, is remarkable,
and should receive the attention of the Registrar-General
without delay.

From the *Lancet*, 1890, vol. i., p. 1440 :—

"UNCERTIFIED CAUSES OF DEATH IN ENGLAND.

" Considering the general progress that has been made
in public health during the last twenty years, it is seriously
to be regretted that this matter of unknown and uncertified causes of death has been practically left untouched,
and its settlement is, therefore, more urgently needed

now than when so often pressed upon the public notice
by the late Dr. William Farr during his connection with
the Registrar-General's department."

The Parliamentary Committee above referred to
omitted an unexampled opportunity of inquiring into
the facts of premature burial. They could have summoned pathologists, who had made trance and catalepsy
a subject of close and searching investigation, as well as
physicians, who, in their practice, have been called in to
decide upon cases of apparent death, and of witnesses
up and down the country who know of such cases, and
others who have met with narrow escapes from these
horrible mishaps. Instead of taking this reasonable
course of procedure, the Committee contented themselves by examining two or three medical men, who had
been summoned to give evidence upon the irregularities
of death-certification only, and whose negative and
apathetic replies showed either that the subject had
never engaged their attention, or that they were unwilling to charge any member of the profession with a
fault so ruinous to his professional reputation as to be
unable to discriminate between the living and the dead.
No questions were submitted to the witnesses as to the
signs of death, the characteristics of catalepsy, trance,
asphyxia, syncope, etc., or how to distinguish these from
death, or as to the submission of tests in doubtful cases
in order to ascertain the fact of death. Indeed, it may
be observed that the investigation regarding a most
vital point connected with death-certification appears to
have entirely escaped the notice of this tribunal. As a
specimen of the proceedings under this head are the
following (" Report," p. 116)—Mr. John Tatham, M.A.,

M.D., being under examination by the chairman, Sir Walter Foster, M.D.

Q. 2112.—Have you ever had any instances within your knowledge, or brought to your notice, of cases where persons have been buried alive?—Never.

Q. 2113—Do you think such cases occur frequently?—I have no means of knowing.

Q. 2114—Supposing the public think they do sometimes, your methods (of medical death-certification) would be a great barrier to anything like that?—Yes.

Q. 2115—The doctor's examination and identification of the body would enable them to detect in many instances if such an occurrence was likely to take place?—I think so.

Further questions were asked of the same witness by Dr. Farquharson.

Q. 2178—You do not believe in people being buried alive?—I do not think that occurs in Manchester.

Q. 2179—Do you think it occurs anywhere?—I do not know.

Q. 2180—We read occasionally very horrifying descriptions of bodies having been found to have turned in their coffins. How do you explain that?—I am not able to explain it.

A correspondent of the *Undertakers' and Funeral Directors' Journal*, July 22, 1893, p. 92, writes :—

"PREMATURE BURIAL.

"Sir,—The newspapers continue to give us fresh accounts of premature burials. Seeing how frequently cases are heard of (in spite of the exhumations being not one-thousandth per cent. of the interments), the occurrence is probably far more common than is generally supposed. It is, therefore, surprising that medical men have not discovered an infallible evidence of death—whatever the cause of death may be; or a simple means of proving, beyond the possibility of doubt, that life is extinct. Further, the application of such a test should, by law, be made to form part of the certificate of death.—I am, Sir, your obedient servant, "LUX.

"July 3."

17

VERIFICATION OF DEATHS.

"In Paris and the large French towns medical inspectors, called *médecins verificateurs*, are appointed, whose business it is to visit each house where a death occurs, and ascertain that the person is really dead, and that there are no suspicious circumstances connected with his or her decease. More than eighty qualified medical men are employed for this purpose in Paris.

"In the rural districts of France this system is not in. force ; two witnesses making a declaration to a civil officer that a death has taken place, is considered sufficient. The burial is not allowed to take place until at least twenty-four hours after the declaration."—*Blyth: Dictionary of Hygiene and Public Health.*

Dr. Léonce Lénormand, in his admirable work " Des Inhumations Précipitées," p. 140, accuses the *médecins des morts* in France with culpable carelessness in the exercise of their function, which consists in verifying the reality of the death. Instead of making a minute examination of the body to ascertain the fact of death, this writer says they are content (except in cases of death from violence) to merely glance at the body, and immediately to hand the family the necessary authorisation for interment. The inspector knows that if he examined every part of the body, as in duty bound, he would be accused of barbarism and profanation. Those, therefore, who think that premature burial could be prevented in England by means only of a more stringent law of compulsory death-certification, would, if it were carried, find themselves in hardly any better position than at present, where the fact of death is left

to a great extent to the judgment of friends, if the
deceased has any, or to the perfunctory inspection of
the undertaker. It is in France where probably, in
spite of *médecins verificateurs*, more premature burials
occur than in any country in Europe except Turkey,
immediate burial after real or apparent death being
the inexorable rule. Dr. Lénormand attributes the frequency of premature burials in France, first of all, to the
negligence and prejudices of the families of the deceased;
then to the carelessness of the doctors charged by the
State with the inspection of the dead ; and lastly, to the
imperfection of the police regulations.

From the *British Medical Journal*, January 28, 1893,
p. 204. (Special Correspondence, Paris.)

"PREMATURE BURIAL.

"The question whether premature burial occurs, and
how to prevent it, is, notwithstanding the all-absorbing
interest of the Panama question, attracting some attention here. The 'Union Medicale' devotes one of its
feuilletons to it, in which two or three *nouvelles à sensation* are reproduced, and easily proved to be untrue.
Premature burial cannot occur, the writer says, when
a death is duly verified. The 77th Article of the Code
obliges the *officier de l'état civil* to visit the death-bed
and verify every death; but this Article is a dead letter.
The officer in question has neither time nor knowledge
sufficient to put it in practice. In small country places,
rarely any precautions are taken to prevent premature
burials. In more important villages and towns, the
mayors delegate the doctors of the locality to verify
deaths before burial. Throughout the whole of France,

it appears that there are not fifty towns where the death-verifying service is well organised ; and, on an average, there are from twenty thousand to thirty thousand burials without previous verification of death. The declaration of two witnesses is sufficient, who obtain their information from those around the deceased. In Paris, the two mortuaries already in existence—one at the Montmartre Cemetery, the other at Père La Chaise —are rarely used. The bodies of those who die in the streets, from accident or sudden death, are taken there when there is no domicile ; also, those of foreigners who die in lodging-houses. In the course of eighteen months the mortuary of Montmartre received five dead bodies, and Père La Chaise one. In Germany the mortuaries are much used, and every arrangement made is in order that any who come back to life may be able to easily summon help. At Munich, a ring in connection with a bell-cord is put on one of the fingers of the hands of the dead. At Frankfort, similar precautions are taken."

Extracts from " Regulations for the Domiciliary Examination of the Dead in the City of Brussels Civil Government (Medical Service)."

"ARTICLE 1.—The Medical Service of the Civil Government is distributed among the medical heads of divisions, the deputies and chiefs of the Department of Hygiene."

"ARTICLE 5.—No interment can take place except after the decease has been verified by the doctors of the Civil Government by means of a careful and complete examination of the corpse."

This verification, as well as the identity of the person deceased, shall be certified by a *procès-verbal* [statement,

or description, for which a blank is furnished "A"],
which they shall leave at the house of the deceased.

"ARTICLE 8.—They shall notify the officers of the Civil Government, and their superintendents of police, of any infractions of
the regulation provisions which forbid proceeding with autopsy,
moulding [making a cast?], embalmment, or putting in a coffin the
corpse, before the death has been duly ascertained."

"ARTICLE 9.—The verification of the decease of still-born or of
newly-born infants shall exact a most attentive examination on
the part of the examining doctors. They shall indicate in their
report if the infant has died before, during, or after birth ; and, in
the last case, how long it lived after birth."

"ARTICLE 10.—If they doubt the reality of the death, they shall
employ, without delay, every means of recovery that science
suggests under the circumstances. They shall immediately notify
the visiting doctor, and, in every case, shall prepare the *procès-
verbal* of the verification of death only after certainty has been
established, and, if need be, by repeated visits."

"ARTICLE 11.—When a woman has died in a state of advanced
pregnancy, they shall direct the artificial extraction of the infant,
supposed to be yet living ; and, in the lack of an attending doctor,
shall perform it themselves when necessary."

EXAMINATION AND CERTIFICATION OF THE DEAD IN WÜRTEMBURG.

A Royal Decree, entitled " Dienst-Vorschriften für
Leichenhaüser," for the inspection and burial of the
dead, promulgated by the King of Würtemburg,
January 24, 1884, provides for the appointment of
medical inspectors of the highest integrity and qualifica-
tions in every commune, the position being justly
regarded as one of great responsibility.

Immediately after a death, the body must under no
circumstances be interfered with, and must not be
removed from the death-bed until after the authorised

inspection. *Post-mortems* can be made only if the fact of death has been previously clearly established. Precise instructions are laid down, so that the inspector, who is to examine the entire body, may see that the various forms of suspended animation are not certified as actual death. Amongst these are the following :—

"Section ii.—To see that sensibility, pulsation of the heart, neck, temples, and forearm, and the breath, have ceased. That the muscles of the body have lost their elasticity; therefore the limbs are limp, the face sunken, the nose pinched, the eyes sunken, and, when the eye-lids are forcibly opened, they remain so, the lower jaw drops more or less, and drops again when pressed upwards.

"In actual death the body gradually gets colder, beginning with the exposed limbs, and in from ten to sixteen hours the body will be quite cold. The colour of the face becomes ashy pale, and the lips discoloured. The eye loses its brilliancy, and is usually dulled by a covering of dried mucus.

"If all the foregoing symptoms are exhibited, and particularly if the deceased was of an advanced age, or if the death was caused by severe or long illness, which led to the expectation of a fatal result, the fact of death may be safely assumed.

"But, on the other hand, if part of these symptoms are missing, or in cases of pregnancy, or exhaustion in consequence of flooding after confinement, or if death occurs under fits, or in violent outbursts of passion, the possibility of counterfeit-death is to be taken for granted.

"Notwithstanding the existence of all the symptoms (signs of death) before mentioned, the possibility of *apparent* death is not excluded in cases where the death has occurred after syncope, tetanus, suffocation, or in cases of drowning, stroke of lightning, or from a severe fall, or from frost, or in still-born children."

After detailing instructions as to a variety of experiments to ascertain whether the death is actual or apparent, this Royal Decree proceeds :—

"Section viii. — These experiments may, however, not give absolute certainty as to the complete extinction of all life. If, therefore, the slightest doubt remains as to the reality of death, the inspector is to take the necessary precautions for the protection of the deceased, by frequent inspections, and the most careful examinations, and to obtain the assistance of the nearest physician or surgeon, who is to co-operate with him to promote resuscitation. If these attempts prove abortive, he must see that nothing is done which would be detrimental to reanimation, or resumption of life."

Then follow minute instructions how to proceed under the varied circumstances which may have produced the symptoms known as apparent death. *In no case must the burial certificate be handed over by the inspector until he has thoroughly satisfied himself of the presence of unmistakable signs of actual death.*

One cannot help contrasting these carefully considered rules with the lax and haphazard methods of dealing with the dead and apparent dead both in England and in the United States. As a consequence, cases of premature burial in Würtemburg are of very rare occurrence, and sensible people in that country, knowing that the danger of premature burial has been reduced to a minimum, are not consumed by an ever-abiding anxiety as with us, nor is it the custom for testators in Würtemburg to give instructions to their executors for piercing the heart or severing the jugular vein, or some other form of mutilation, as in France, Spain, and other countries, where the risks are so terribly great.

The only case of the danger of premature burial that has come to the author's notice in Würtemburg is related by Bouchut, in his " Signes de la Mort," p. 48 :—

"In the village of Achen, in Würtemburg, Mrs. Eva Meyers, twenty-three years of age, was taken ill during an epidemic. Her condition became rapidly worse, and she apparently died. They put her into a coffin, and carried her from the warm into a cold room, there to await burial, which was to take place at two p.m. on the following day. Shortly after noon on that day, and before the carriers arrived, she awoke and made an effort to rise. Her aunt, who was present, and who believed that a ghost had taken possession of her, took a stick and would have killed her, if she had not been prevented by another woman. Nevertheless, she succeeded in pushing the body back violently into the coffin, after which she indignantly went to her room. The patient remained helplessly in that condition, and would have been buried if the usual hour for the burial had not for some reason been changed. Thus she remained for another twelve hours, when she was able to gather sufficient strength to arise. She still lives, and has paid the charges for her funeral, which were claimed by the clergy, the bell-ringer, and the undertaker."

In the United States the subject of Death-Verification has only recently begun to engage public attention. The following appears to be the only instance in which reasonable, although not altogether adequate, precautions are adopted.

"DOVER, NEW HAMPSHIRE.—CITY ORDINANCES, 1895.

"CHAP. XVII.—VITAL STATISTICS.

" SECTION 3.—Whenever any person shall die within the limits of the city, it shall be the duty of the physician, attending such person, during his or her last sickness, to examine the body of such deceased person before the burial thereof, and to make out a certificate, setting forth, as far as the same may be ascertained, the name, age, colour, sex, nativity, occupation, whether married or single, duration of residence in the city, cause, date, and place of death of such deceased person ; and it shall be the duty of the undertaker, or other person in charge of the burial of such deceased person, to add to such certificate the date and place of burial, and, having duly signed the same, to deposit it with the city clerk, and obtain a permit for burial ; and,in the case of death from any contagious or infectious disease, said certificate shall be made and forwarded immediately; and,in each case of a physician so examining and reporting, he shall receive of the city a fee of one dollar."

"SECTION 4.—Whenever a permit for burial is applied for, in case of death without the attendance of a physician, or it is impossible to obtain a physician's certificate, it shall be the duty of the city physician to make the necessary examination, and to investigate the case, and make and sign a certificate of the probable cause of death ; and,if not satisfied as to the cause and circumstances attending such death, he shall so report to the mayor."

"SECTION 5.—No interment or disinterment of the dead body of any human being, or disposition thereof in any tomb, vault, or cemetery, shall be made within the city without a permit therefor, granted as aforesaid, nor otherwise than in accordance with such permit.

"No undertaker, superintendent of cemetery, or other person shall assist in, assent to, or allow any such interment, or disinterment, to be made, until such permit has been given as aforesaid. . . . .

"Any person violating any of the provisions of this chapter shall be fined not less than ten nor more than twenty dollars."

Mr. A. Braxton Hicks, Barrister-at-Law, and Coroner for London and Surrey, states that—

"The giving of certificates of death, and the registration of deaths, is regulated by 37 and 38 Vict. c. 88, called the Registration of Births and Deaths Act, its object being to provide a proper and accurate registration of births and deaths, with the causes of the latter.

"In case of the death of any person who has been attended during his last illness by a registered medical practitioner, that practitioner shall sign and give to some person, required by this Act to give information concerning the death, a certificate stating, to the best of his knowledge and belief, the cause of death.

"No certificate given by an unregistered medical man can be registered, and any person who covers an unregistered medical man by giving a certificate, or lending his name to the giving of a certificate by an unregistered medical man, is guilty of *unprofessional conduct*, as defined. by the Medical Council."— *Hints to Medical Men concerning the granting of Certificates of Death.*

A DOCTOR FOR THE DEAD.

Dr. J. Brindley James, in a communication to the *Medical Times*, May 23, 1896, pp. 355-356, calls attention to the insufficient safeguards against premature burial under the present system of death-certification, and observes—" The dread possibility of premature interment ever hangs like a gloomy sword of Damocles over all our heads, and fearful indeed is the authentic record of persons buried alive, who have recovered consciousness ; too late, alas ! to be rescued from their frightful dungeon. How often does our overworked—we do not say careless—practitioner sign the death-certificate of a patient whose death-bed he did not attend—whose corpse he has not visited ? And even assuming him to have done so, and conscientiously too, in how many of the fearful cases above alluded to have not these formalities proved insufficient, clearly suggesting the advisability of a specialist, experienced in *post-mortem* inspection, solely sanctioning interment in all cases." And Dr. Frederick Graves, writing in the same journal of July 18, 1896, says :—

" I have recently heard of a case which illustrates the utility of a medical examination before burial. A soldier in the German army, during the forced march on Paris, became unconscious, with five others, from sunstroke, and the six were put aside for burial by their comrades, when the timely examination of the army surgeon prevented premature burial of the person referred to, who is alive and well at the present time."

The *Daily Chronicle*, London, September 16, 1895, in a leading article on the danger of premature burial, says :—" The truth is, the whole system of certifying for burial needs to be reconsidered and reformed, and that for other reasons than the danger of entombment before life is extinct.   We do not want a coroner's inquest, with its jury, for every death ; but the doctors should be compelled, under severe penalties, to discover the certain sign of death before they authorise the burial, and to know the cause of death in every case.   We trust now too much to individuals in a generally trustworthy profession, who may not reach the high general standard of their class, or may grow listless through the indifference wrought by use and wont, or who think they can detect the *rigor mortis* at a glance, never having seen the severest form of catalepsy.   There would be no difficulty in getting Parliament to pass a more stringent regulation for death-certificates without much discussion, and there is no reason why Sir Matthew White Ridley should not turn his attention to the matter, and, with such medical advice as the Health Department of the Local Government Board will be pleased to lend him, propose a necessary little bill to the House of Commons next February."

The following letter by a German resident in England appeared in the *Times* of September 20, 1895 :—

"BURIED ALIVE.

"Sir,—As this important subject appears to be arresting the attention of the public in England, may I venture to state the law as to the examination of corpses in my own country? In a copy of the official regulations in Würtemburg for the inspection of dead bodies ('Dienst-Vorschriften für Leichenhaüser in Würtemburg, 1882.' Stuttgart, W. Kohlhammer), I find the following :—

"'No corpse must be interfered with before the arrival of the inspector, who is expected to pay several visits before granting the death-certificate, which he alone is authorised to do. In cases of death from infectious disease the body must be removed to a mortuary, where it is carefully watched.'

"These inspectors are highly qualified, State-appointed physicians, but, as if to show the uncertainty of all this care and experience, as we see by the researches of Dormodoff, Hufeland, Hartmann, and others, as well as by the reports of startling cases in the press, those medically and officially declared to be dead do occasionally come to life before burial. This is a state of things unworthy of the civilisation and humanity of which we are proud.

"Medical examination, not being infallible when carried out at its best, must be very unreliable when performed in a careless manner.

"A safer plan would be to send every supposed corpse to a mortuary, there to remain until decomposition manifests itself. As a German I should be afraid to die in England (excuse the paradox) for fear of being buried alive.

"P. P.

"Forest Hill, September 17."

CHAPTER XIX.

THE learned Dr. Vigné, of Rouen, who won the respect of his fellow-citizens during a long and honourable career, was for many years engaged in the study of this question, and published the result of his researches shortly before his death. Convinced that the resources of science were insufficient to distinguish real from apparent death, he left testamentary instructions to provide against his own premature burial. ("Des Inhumations Précipitées, p. 83," by Lénormand.)

Dr. Winslow, a French physician, who had on two different occasions very nearly fallen a victim to premature burial, having been laid out for dead, chose for the subject of his thesis before the Paris Faculty of Medicine, " Les moyens les plus propres à reconnaître la réalité de la mort." Dr. Winslow may be said to have been the pioneer of a movement in France for exposing the danger of, and educating the public into the necessity of reforms in, the mode of treating the apparent dead ; and, although his efforts and warnings were as of one crying in the wilderness or amongst an apathetic people, with a legislature apparently uninfluenced either by facts or by reason, they were never relaxed. Numerous writers have since confirmed the truth of Dr. Winslow's contention by facts within their own experience, and it is believed that legislation in France cannot be much longer delayed.

That the risk of premature burial is not an imaginary one, as recently declared by a leading London medical journal, has been shown by the citation in this volume of cases of death-like trance which have baffled the ablest of medical experts; also the instances of numerous narrow escapes from this terrible occurrence, and of others where the victims were suffocated before timely aid could be obtained, most of which are drawn from medical sources, and some from the columns of the said sceptical journal. The painful reality is also shown by the multitude of preventive measures suggested by medical authorities, and by the ingenious contrivances of those who have made this distressing subject one of patient and laborious research. Several of the remedies suggested for adoption in cataleptic cases are really homicidal, or seriously mutilative; many of them are impracticable, and have been shown by Hufeland, Lénormand, Richardson, Hartmann, Bouchut, Fletcher, and Gannal to be delusive. The merits and demerits of some of these methods might be inquired into by the appointment of a Parliamentary Committee, or a Royal Commission, as a supplement to that appointed in 1893, by Mr. Asquith, on Death-Certification.

CUTANEOUS EXCITATION.

Dr. James Curry, F.R.S., in his "Observations on Apparent Death," pp. 56, 57, says, concerning the application of stimulants to the skin :—

"To assist in rousing the activity of the vital principle, it has been customary to apply various stimulating matters to different parts of the body. But as some of these applications are in themselves positively hurtful, and the others serviceable only according to the time and manner of their employment, it will be proper to consider them particularly.

"The application of all such matters in cases of apparent death is founded upon the supposition that the skin still retains sensibility enough to be affected by them. It is well known, however, that even during life the skin loses sensibility in proportion as it is deprived of heat, and does not recover it again until the natural degree of warmth be restored. Previous to the restoration of heat, therefore, to a drowned body, all stimulating applications are useless, and, so far as they interfere with the other measures, are also prejudicial."

Several writers, besides Dr. Winslow, whose views on premature burial are cited in this volume, have themselves been the victims of hasty and erroneous medical diagnosis ; and having had narrow escapes of premature burial, their experience has prompted them to take a deep interest in the subject, with the determination to do what they could to enlighten and safeguard the public from so terrible a danger. In other cases, members of their families have been the unhappy victims of mistaken certificates. Mr. George T. Angell, the editor of "*Dumb Animals*," Boston, U.S., whose father was pronounced by his physician dead, and returned to consciousness after preparations for the funeral had been made, has repeatedly alluded to the subject in his paper, and published preventive suggestions at various times, including the following from a physician :—

"When I arrived, the man had been dead twenty-four hours. I empanelled a jury ; the family of the deceased testified to the extent of their knowledge ; but I was unable to find he had any disease sufficient to kill him. I looked at the body and examined it carefully. Then I lighted a match, and applied it to the end of one of the fingers of the corpse. Immediately a blister formed. I had the man put back into his bed, applied various restoratives, and to-day he is alive and well.

"That is the test. Do you see the philosophy of it? If you are alive, you cannot burn your hand without raising a blister. Nature, in the effort to protect the inner tissues, throws a covering of water, a non-conductor of heat, between the fire and the flesh. If you were dead, and flames should come in contact with any part of your body, no blister would appear, and the flesh would be burned.

"All you have to do is to apply a match to any part of the supposed corpse.   If life remains, however little, a blister will at once form."

The test, like the following one, is deceptive, because life may be so torpid and inactive as to be unable to respond to the irritation of heat, or even to the application of red hot irons.

THE BLISTER TEST.

The *British Medical Journal*, January 18, 1896, p. 180, under the head of "Living or Dead?" prints the following communication concerning this test :—

"Sir,—Burial alive, though of exceedingly rare occurrence, sometimes does happen, and calls for increased attention to the means of detecting with certainty the presence of vitality, however feeble. The ordinary means of deciding the vital question are known to all persons.   Auscultation may detect the enfeebled heartbeat, while the electric battery can elicit any existing muscular contractility.   Conditions of trance are occasionally almost mystical in their profundity (Brahmin trance), and a simple and ready-to-hand test to decide whether death has occurred is of prime importance. We can ascertain whether or not life still lingers in uncertain cases by applying (say) to the back of the forearm a small stream of boiling water directly from the kettle. If life is present, the boiling water will soon and unfailingly raise a blister where applied, and the blister will contain fluid, the serum of the blood. The production of the serum blister being essentially a vital process, its production or non-production becomes an infallible test, and determines the question. This test, not generally known, should be widely proclaimed.

"J. MILFORD BARNETT, M.D., Edin.

"Belfast, January 11, 1896.'

This test has frequently failed, and should not be relied upon.

AUSCULTATION.

The stethoscope, which is regarded by many medical practitioners as an infallible means of preventing premature burial, has proved a broken reed in hundreds of cases, and can be of use only when applied with other tests. Dr. Roger S. Chew, of Calcutta, writes to me, February, 1896 :—

"The *British Medical Journal* (September 28, 1895) tells us that the careful use of the stethoscope will enable a medical man to distinguish a living from a dead body. Auscultation may give startling results, and the body yet be absolutely dead. I recollect an instance of death from cobra-bite, when, though decomposition had set in, the relatives refused to believe she was dead, because one of them declared that, though he did not see her chest rise and fall, he had distinctly heard her sigh. A medical man was called in, applied the stethoscope over her thorax, and declared he could hear sounds from her lungs, and a peculiar '*sough*,' '*sough*' towards the

18

apex of the heart. So far he was right, but, as the girl had already been dead some fourteen hours, and the weather was warm, the sounds he heard were those of the escape of the putrefactive gases bubbling upward and unable to find exit, as her mouth was closed with a chin-bandage, and her nostrils plugged with mucus. To convince the parents that the girl was really dead, I offered to perform artificial respiration, to which end I untied the bandage, prized open her jaws, and pressed *heavily* on her thorax, when some of the imprisoned gases escaped, emitting an abominable odour that brought conviction of the girl being beyond all hope.

"In another case, that of my son, aged two years, after a series of brain symptoms and severe clonic convulsions preceding an outbreak of confluent small-pox, the stethoscope told me and a medical friend who was present that my little boy had ceased to exist; but a liberal application of ice to his head and cardiac region, together with violent friction and artificial respiration vigorously employed for *forty* minutes, restored the child to me, and I thanked God that I had refused to accept the evidence of the stethoscope as final."

ELECTRICITY.

The application of the electric current is a powerful restorative agent in cases of suspended animation, if judiciously applied. Struve in his essay, "Suspended Animation," p. 151, under the head of "Apparent death from a fall," says:—"A girl, three years of age, fell from a window two stories high upon the pavement. Though she was considered as lifeless,

Mr. Squires, a natural philosopher, applied electricity. Almost twenty minutes elapsed before the shocks produced any effect. At last when some of the electric force pervaded the breast he observed a slight motion of the heart. The child soon after began to breathe and groan with great difficulty, and after some minutes a vomiting ensued. For a few days the patient remained in a state of stupefaction, but in the course of a week she was perfectly restored to health."

Referring to the subject of premature burial, Dr. W. S. Hedley, writing to the *Lancet*, October 5, 1895, says:—"Forty years ago the subject was investigated by Crimotel, twenty years later by Rosenthal, and more recently by Onimus. It seems safe to say that in no disease, certainly in none of those conditions usually enumerated as likely to be mistaken for death, is galvanic and faradaic excitability abolished in every muscle of the body. On the other hand, electro-muscular contractility disappears in all the muscles within a few hours after death (generally ninety minutes to three hours, according to Rosenthal), its persistence varying to some extent with the particular muscle examined (1), and with the mode of death (2). Therefore, if electro-muscular contractility be present in any muscle, it means life or death only a few hours before. It is clear that no interment or *post-mortem* examination ought to take place so long as there is any flicker of electric excitability. To me it seems almost equally obvious that in all doubtful cases, sometimes in sudden death, and often to allay the anxiety of friends, this test ought to be applied, and applied by one who is accustomed to handle electric currents for purposes of diagnosis."

The *Medical Record*, New York, March 30, 1895, contains the following :—" In a case reported by M. D'Arsonval, a man was struck with a current of four thousand five hundred volts. The current entered at his hand and issued at his back. Half an hour or more elapsed before any attempts at resuscitation were made, but, on artificial respiration being practised on Silvester's method, recovery took place. Dr. Donnellan reports a case of the passage of a current of one thousand volts through a man, which instantly caused coma, dilated pupils, pallor of the face, and sweating; delirium and tonic, alternating with clonic, spasms followed. The pulse was eighty. The respiration, at first stertorous, passed into the Cheyne-Stokes type. After the injection, first of morphia, and then of strychnia, the patient fell into a deep sleep, from which he awoke convalescent.—*Central-blatt für die medicinischen Wissenschaften.*"

The apparatus for applying electrical currents, long used by the Humane Society for restoration of the drowned, might with advantage be kept at public mortuaries, for use in cases of apparent death due to other causes, where decomposition has not manifested itself. The Weather Bureau at Washington advises those who are in the neighbourhood of persons struck by lightning to make immediate efforts to restore consciousness, because the effect of lightning is to suspend animation rather than to produce death. Respiration and circulation should be stimulated, and the usual remedies for relief in such cases should be administered for at least an hour before giving up the victim as dead.

Dr. Moore Russell Fletcher says :—"When persons without pulse or breathing are found in bed, in the field, or elsewhere, treat them in such manner as will restore from stroke of lightning, paralysis, or suspended animation from catalepsy, trance, or somnambulism, and continue the treatment until resuscitation rewards the exertions, or decomposition is evident."—*Suspended Animation, pp. 7, 8.*

HYPODERMIC INJECTIONS.

Mr. E. E. Carpmael, of the Medical Department, Berkeley University, U.S.A., recommends, in the *Morning Post*, London, September 19, 1895, the injection of strychnine in "a supposed corpse ;" while " Medicus," in the *Daily Chronicle*, September 17, 1895, considers that *post-mortems* "would be to the advantage of the patient, to his relations, to science, and the community at large." No doubt either of these plans would prevent live sepulture, by killing the cataleptic subject ; while " M.R.C.S.," in *Morning Post*, September 20, says :— " Obviously the simplest and best proof of death is putrefaction—shown chiefly by the discolouration of the abdomen."

A correspondent in the *English Mechanic*, October 25, 1895, says :—" I have long advised hypodermic injection of morphia before placing in coffin for burial. *Ex hypothesi*, the vital spark is not supposed to have expired, and the circulatory system not finally stopped. Hence the hypodermic injection cannot be futile."

A medical correspondent writing from Dresden, August 18, 1895, sends me the following as showing the value of

ARTIFICIAL RESPIRATION.

"Major J. H. Patzki, Surgeon, U.S. Army, reports that in 1882, at St. Augustine, Florida, a lady patient of his had an attack of tetanus, caused by a scratch upon her foot by a nail while bathing. The convulsive symptoms commenced in the muscles of the face, and increased in violence in spite of energetic treatment, until the fifth day, when the respiratory muscles became involved. The breathing was completely suspended by the spasmodic action, and the radial and carotid pulse ceased. The cardiac sounds became utterly inaudible to careful stethoscopic examination repeatedly employed. The lady assumed all the appearances of death, and there was *rigor mortis*, the result of muscular spasm. Artificial respiration was resorted to, but not until after the expiration of eighteen minutes did the first faint efforts of respiration, and a feeble action of the heart, become perceptible. Artificial respiration was continued for an hour afterwards, and the life of the patient was saved, although the muscular spasms continued to some extent for six days.

"This case is instructive in showing that tetanus, when it involves the chest, may produce a state of apparent death, by interfering with the respiratory and cardiac functions ; and that artificial respiration, if persistently employed, may rescue patients so affected from the perils of apparent death."

Dr. John Oswald, in "Suspended Animal Life," Philadelphia, 1802, p. 65, says :—"The books of authors on this subject are replete with criteria to judge of the existence or non-existence of the vital principle. It is not necessary to take a separate view of the propriety or impropriety of adopting any of these ambiguous signs, when we have the accomplishment of so great an end as that of restoring suspended life! Our exertions should never be influenced by any of them, but continued with ardour and unremitted attention for a length of time. It would be more happy for our unfortunate patients, and a source of greater satisfaction to ourselves, were they expunged altogether. They are all fallacious to a certain degree, and ought never to have the smallest influence on the propriety or impropriety of persevering in our attempts to revive the latent spark; for it is an unfortunate fact, in consequence of an ignorant confidence placed in them, that persons who might have been restored to life, to their friends, and to society have been consigned to the grave. . . . This important subject has been anxiously investigated by philosophers, to discover a just criterion of judging with more certainty in these cases whether life is extinct, and our patient a mass of dead matter, or whether, by our perseverance, he may not be again recovered. The most indubitable sign is allowed to be putrefaction of the body, or disorganisation of the fibre."

The following extracts from an instructive but apparently forgotten article in Dickens' "All the Year Round," July, 1869 (à propos of a pamphlet, "Lettre sur la Mort Apparente, les Conséquences Réelles des Inhumations Précipitées, et le Temps pendant lequel peut persister l'Aptitude à être rappellé à la Vie," by the late regretted Dr. Charles Londe), afford valuable suggestions :—

"Suffocation by foul air and mephitical gas is not a rare form of death in the United Kingdom. It is possible that suspended animation may now and then have been mistaken for the absolute extinction of life. Dr. Londe gives an instructive case to the purpose. At the extremity of a large grocer's shop, a close, narrow corner, or rather hole, was the sleeping-place of the shopman who managed the night sale till the shop was closed, and who opened the shutters at four in the morning. On the 16th of January, 1825, there were loud knocks at the grocer's door. As nobody stirred to open it, the grocer rose himself, grumbling at the shopman's laziness, and, proceeding to his sleeping-hole to scold him, he found him motionless in bed, completely deprived of consciousness. Terror-struck by the idea of sudden death, he immediately sent in search of a doctor, who suspected a case of asphyxia by mephitism. His suspicions were confirmed by the sight of a night-lamp, which had gone out, although supplied with oil and wick, and by a portable stove containing the remains of charcoal partly reduced to ashes. In spite of a severe frost, he immediately had the patient taken into the open air, and kept on a chair in a position as nearly vertical as possible. The limbs of the sufferer hung loose and drooping, the pupils were motionless, with no trace either of breathing or pulsation of the heart or arteries ; in short, there were all the signs of death. The most approved modes of restoring animation were persisted in for a long while without success. At last, about three in the afternoon, —that is, after *eleven hours'* continued exertion,—a slight movement was heard in the region of the heart. A few hours afterwards the patient opened his eyes, regained consciousness, and was able to converse with the spectators attracted by his resurrection. Dr. Londe draws the same conclusions as before—namely, that persons suffocated by mephitism are not unfrequently buried when they might be saved."

"We have had cholera in Great Britain, and we may have it again. At such trying times, if ever, hurried interments are not merely excusable, but almost unavoidable. Nevertheless, one of the peculiarities of that fearful disease is to bring on some of the symptoms of death—the prostration, the coldness, and the dull livid hues—long before life has taken its departure. Now, Dr. Londe states, as an acknowledged fact, that patients pronounced dead of cholera have been repeatedly seen to move one or more of their limbs after death. While M. Trachez (who had been sent to Poland to study the cholera) was opening a subject in the dead-house of the Bagatelle Hospital, in Warsaw, he saw another body (that of a woman of fifty, who had died in two days, having her eyes still bright, her joints supple, but the whole surface extremely cold) which vividly moved its left foot ten or twelve times in the course of an hour. Afterwards, the right foot participated in the same movement, but very feebly. M. Trachez sent for Mr. Searle, an English surgeon, to direct his attention to the phenomenon. Mr. Searle *had often remarked it.* The woman, nevertheless, was left in the dissecting-room, and thence taken to the cemetery. Several other medical men stated that they had made similar observations. From which M. Trachez draws the inference : 'It is allowable to think that many cholera patients have been buried alive.' "

" Dr. Veyrat, attached to the Bath Establishment, Aix, Savoy, was sent for to La Roche (Department of the Yonne), to visit a cholera patient, Thérèse X., who had lost all the members of her family by the same disease. He found her in a complete state of asphyxia. He opened a vein ; not a drop of blood flowed. He applied leeches ; they bit, and immediately loosed their hold. He covered the body with stimulant applications, and went to take a little rest, requesting to be called if the patient manifested any signs of life. The night and next day passed without any change. While making preparations for the burial, they noticed a little blood oozing out of the leech-bites. Dr. Veyrat, informed of the circumstance, entered the chamber just as the nurse was about to wrap the corpse in its winding-sheet. Suddenly a rattling noise issued from Thérèse's chest. She opened her eyes, and in a hollow voice said to the nurse : 'What are you doing here? I am not dead. Get away with you.' She recovered, and felt no other inconvenience than a deafness, which lasted about two months."

"Exposure to cold may also induce a suspension of vitality liable to be mistaken for actual death. This year the French Senate has again received several petitions relative to premature interments. . . . And, considering the length of time that trances, catalepsies, lethargies, and cases of suspended animation have been known occasionally to continue, it is scarcely, in England, less interesting to us, though public feeling, which is only an expression of natural affection, approves, and indeed almost compels, a longer delay. The attention of the French Government being once more directed to the subject, there is little doubt that all reasonable grounds for fear will be removed.[1]

" The petitioners have requested, as a precaution, that all burials for the future should, in the first instance, be only provisional. Before filling a grave, a communication is to be made between the coffin and the upper atmosphere by means of a respiratory tube ; and the grave is not to be finally closed until all hope of life is abandoned. These precautions, it will be seen at once, however good in theory, are scarcely practicable. Others have demanded the general establishment of mortuary chambers, or dead-houses, like those in Germany. And not only the petitioners, but several senators, seem to consider that measure the full solution of the problem. Article 77 of the Civil Code prescribes a delay of twenty-four hours only, which appears to them to be insufficient, since, they urge, it admits the certainty that death has taken place only after putrefactive decomposition has set in. Now, a much longer time than twenty-four hours may elapse before that decomposition manifests itself. Deposit, therefore, your dead in a mortuary chapel, until you are perfectly sure, from the evidence of your senses, that life is utterly and hopelessly extinct.

"When Article 77 of the Civil Code was under discussion by the Council of State, Fourcroy added : 'It shall be specified that the civil officer be assisted by an officier de santé (a medical man of inferior rank to a doctor of medicine) ; because

---

[1] Alas for the futility of human expectations of reform when left to the initiation of Governments—this was written twenty-seven years ago, and nothing has been done to remedy the evil !

there are cases in which it is difficult to make certain that death has actually occurred, without a thorough knowledge of its symptoms, and because there are tolerably numerous examples to prove that people *have* been buried alive.'

"In Paris, especially since Baron Hausmann's administration, Article 77 has been strictly fulfilled; but the same exactitude cannot be expected in out-of-the-way nooks and corners of the country, where a doctor cannot always be found at a minute's warning, to declare whether death be real or apparent only. It is clear that the Legislature has hit upon the sole indisputable practical solution; the difficulty lies in its rigorous and efficient application.

"It has been judiciously remarked that it would be a good plan to spread the knowledge of the sure and certain characteristics which enable us to distinguish every form of lethargy from real death. It cannot be denied that at the present epoch the utmost pains are taken to popularise every kind of knowledge. Nevertheless, it makes slow way through the jungles of prejudice and vulgar error. Not long ago it was over and over again asserted that an infallible mode of ascertaining whether a person was dead or not was to inflict a burn on the sole of the foot. If a blister full of water resulted, the individual was not dead; if the contrary happened, there was no further hope. This error was unhesitatingly accepted as an item of the popular creed.

"The Council of Hygiene, applied to by the Government, indicated putrefaction and cadaverous rigidity as infallible signs of actual death. In respect to the first—putrefaction—a professional man is not likely to make a mistake; but nothing is more possible than for non-professionals to confound hospital rottenness (gangrene) with true *post-mortem* putrefaction. M. de Parville declines to admit it as a test adapted for popular application. Moreover, in winter, the time required for putrefaction to manifest itself is extremely uncertain.

"The cadaverous rigidity—the stiffness of a corpse—offers an excellent mode of verifying death; but its value and importance are not yet appreciable by everybody, or by the first comer. Cadaverous rigidity occurs a few hours after death; the limbs, hitherto supple, stiffen; and it requires a certain effort to make them bend. But when once the faculty of bending a joint is forcibly restored—to the arm, for instance—it will not stiffen again, but will retain its suppleness. If the death be real, the rigidity is overcome once for all. But if the death be only apparent, the limbs quickly resume, with a sudden and jerking movement, the contracted position which they previously occupied. The stiffness begins at the top,—the head and neck,—and descends gradually to the trunk.

"These characteristics are very clearly marked ; but they must be caught in the fact, and at the moment of their appearance, because, after a time of variable duration, they disappear. The contraction of the members no longer exists, and the suppleness of the joints returns. Many other symptoms might be added to the above ; but they demand still greater clearness of perception, more extended professional knowledge, and more practised habits of observation.

"Although the French Government is anxious to enforce throughout the whole empire the rules carried out in Paris, it is to be feared that great difficulties lie in the way. The verification of deaths on so enormous a scale, with strict minuteness, is almost impracticable. But, even if it were not, many timid persons would say : ' Who is to assure us of the correctness of the doctor's observations ? Unfortunately, too many terrible examples of their fallibility are on record. The professional man is pressed for time. He pays a passing visit ; gives a hurried glance ; and a fatal mistake is so easily made !' Public opinion will not be reassured until you can show, every time a death occurs, an irrefutable demonstration that life has departed.

"M. de Parville now announces the possibility of this great desideratum. He professes to place in any one's hands a self-acting apparatus which would declare not only whether the death be real, but *would leave in the hands of the experimenter a written proof of the reality of the death.* The scheme is this : It is well known that atropine — the active principle of *belladonna*— possesses the property of considerably dilating the pupil of the eye. Oculists constantly make use of it when they want to perform an operation, or to examine the interior of the eye. Now, M. le Docteur Bouchut has shown that atropine has no action on the pupil when death is real. In a state of lethargy, the pupil, under the influence of a few drops of atropine, dilates in the course of a few minutes ; the dilatation also takes place a few instants after death ; but it ceases absolutely in a quarter of an hour, or half an hour at the very longest ; consequently the enlargement of the pupil is a certain sign that death is only apparent.

" This premised, imagine a little camera obscura, scarcely so big as an opera-glass, containing a slip of photographic paper. which is kept unrolling for five-and-twenty or thirty minutes by means of clockwork. This apparatus, placed a short distance in front of the dead person's eye, will depict on the paper the pupil of the eye, which will have been previously moistened with a few drops of atropine. It is evident that, as the paper slides before the eye of the corpse, if the pupil dilate, its photographic image will be dilated ; if, on the contrary, it remains unchanged, the image will retain its original size. An inspection of the paper then enables the experimenter to read upon it whether the death is real or apparent only. This sort of declaration can be handed to the civil officer, who will give a permit to bury in return.

" By this simple method a hasty or careless certificate of death becomes impossible. The instrument applies the test, and counts the minutes. The doctor and the civil officer are relieved from further responsibility. The paper gives evidence that the verification has actually and carefully been made ; for suppose that half an hour is required to produce a test that can be relied on, the length of the strip of paper unrolled marks the time during which the experiment has been continued. An apparatus of the kind might be placed in the hands of the minister or one of the notables of every parish. Such a system would silence the apprehensions of the most timid ; fears—natural enough—would disappear, and the world would be shocked by no fresh cases of premature burial."

The authors have not heard whether this ingenious contrivance had been put into practice, or with what result.

Various prizes have been offered, and awards made, by scientific and medical societies, but, with one exception, the so-called proofs of death for which the awards have been given are deemed unsatisfactory. The most notable of the prizes is that of the Marquis d'Ourches, who by his will bequeathed the sum of twenty thousand francs to be given to the author of the discovery of a simple and common means of recognising beyond doubt the absolute signs of death, by such a test as could be adopted by poor villagers without technical instruction. The Marquis d'Ourches left also a prize of five thousand francs for a similar discovery, but requiring the intervention of an expert. M. Pierre Manni, Professor at the University of Rome, offered a prize, which was awarded to Dr. E. Bouchut, in 1846. And M. Dusgate, by will, dated January 11, 1872, bequeathed to the French Academy of Sciences a sufficient sum in French *Rentes*, to found a quinquennial prize of two thousand five hundred francs to the author of the best work on the diagnostic signs of death, and the means of preventing premature interments. A decree of November 27, 1874, authorised the Academy to accept this legacy.

Dr. Gowers, on " Diseases of the Nervous System," vol. ii., p. 1037, says :—" In cases of ' death-trance,' in which no sign of vitality can be recognised, the presence of life may be ascertained (1) by the absence of any sign of decomposition ; (2) by the normal appearance of the *fundus oculi* as seen with the ophthalmoscope ; (3) by the persistence of the excitability of the muscles to electricity. This excitability disappears in three hours after actual death. In a case observed by Rosenthal, thirty hours after supposed death, the muscles were still excitable, and the patient awoke."

The *British Medical Journal*, January 21, 1893, p. 145,

reports, through its Paris correspondent, the first award.
" The Académie des Sciences proposed as the subject for
the Dusgate Prize for 1890, ' The Signs of Death, and the
Means of Preventing Premature Burial.' The prize has
been awarded to Dr. Maze, who considers that putrefaction is the only certain sign. He urges that the
deaths should be certified by medical men on oath ;
also that in every cemetery there should be a mortuary
where dead bodies can be deposited, and that burial
should take place only when putrefactive changes set in.
Cremation should be adopted."

CHAPTER XX.

AMONGST the numerous suggestions made by correspondents in the press with a view of preventing live sepulture, none has been more frequently put forward than that of cremation. Sir Henry Thompson, the president of the Cremation Society of England, in the second edition of his admirable volume, " Modern Cremation : Its History and Practice," p. 41, observes :—
" There is a source of very painful dread,—as I have reason to know,—little talked of, it is true, but keenly felt by many persons at some time or another, the horror of which to some is inexpressible. It is the dread of a premature burial—the fear lest some deep trance should be mistaken for death, and that the awakening should take place too late. Happily such occurrences must be exceedingly rare, especially in this country, where the interval between death and burial is considerable, and the fear is almost a groundless one. Still, the conviction that such a fate is possible—which cannot be altogether denied—will always be a source of severe trial to some. With cremation no such catastrophe could ever occur ; and the completeness of a properly-conducted process would render death instantaneous and painless if by any unhappy chance an individual so circumstanced were submitted to it. But the guarantee against this danger would be doubled,

since inspection of the entire body must of necessity immediately precede the act of cremation, no such inspection being possible under the present system." While agreeing with this distinguished authority as to the advantages of cremation from the sanitary and æsthetic point of view, which he dwells upon in the treatise referred to, and admitting that a certain amount of protection against live burial is obtainable by means of the dual medical inspection, we cannot agree that this protection is absolute. Cases of trance are on record where some half a dozen doctors, after careful examinations, have pronounced a cataleptic patient to be dead, and the patient, in defiance of their united opinion, has recovered consciousness, and been restored to health.

Dr. Franz Hartmann, in his "Premature Burial," quotes the two following cases amongst many others:—

"Madame de P——, aged eighteen years, and subject to hysteria, apparently died, and for forty hours she presented all the signs of real death. All possible means of restoring her to life were taken, but proved of no avail. *Five physicians of Lyons were called in, and they finally agreed, positively, that the lady was really dead.* The funeral preparations were made; but owing to the supplications of a sister of the deceased the burial was delayed, when after a while the patient recovered. She said that she had been all the time aware of all that was going on, without being able to give a sign, and without even being desirous of attempting it." (F. Kempner, p. 38.)

"In 1842 a remarkable affair occupied the attention of the court at the city of Nantes. A man apparently died, and *his death was certified to both by the attending physicians and the medical inspector;* he was put into a coffin, and the religious cere-monies were performed in good style. At the end of the funeral service, and as he was about to be buried, he awoke from his trance. The clergy and the undertakers sent in their accounts for

19

the funeral expenses ; but he refused to pay them, giving as his reason that he had not ordered them ; whereupon he was sued for the money." (F. Kempner, p. 39.)

Neither can we share the optimistic views of Sir Henry Thompson as to the rarity of premature interment. The results of searching and independent inquiries and study in various countries by each of the authors of this treatise all point the other way, and the various authorities whose names and opinions are cited elsewhere in this volume confess their astonishment at the number of cases brought to light during their investigations. The Rev. H. R. Haweis also, in his work "Ashes to Ashes : A Cremation Prelude" (London, 1895, now out of print), advocates cremation on the ground of preventing living burial, and quotes several cases of persons buried while in a state of trance. During a discussion on the merits and demerits of cremation in the *Birmingham Gazette*, September 17, 1895, Lieutenant-General Phelps, an able and judicious observer, advocated cremation for similar reasons, and said that "the use of a crematorium would entirely prevent that ghastly accident, the burial of the living. There is no room to doubt that this frightful catastrophe is of continual occurrence. The phenomena of trance are little understood, and a certificate of death is held by most of us to justify the burial of the 'corpse,' dead or alive. Those of us who object to the risk of being buried alive should do all in our power to promote the success of this sanitary contrivance for disposing of our dead."

The writer of the following communication, which appeared in the *Sunday Times*, September 6, 1896, has

substantial reasons for preferring cremation to the
risks of burial :—

"BURIAL DANGER AND ITS PREVENTION.

"Madam,—When I was about five years old, my paternal
home was one day plunged into a state of great consternation,
through the sudden apparent death of my father, who had been
sitting up during a part of the previous night occupied with some
literary work, without a fire (it was in January), which brought on
a death-like numbness, in which he was found the next morning.
The family doctor, who was sent for at once, declared life to be
extinct, but said he could not tell the cause of death until after the
opening of the dead body. My mother, however, who did not see
any reason why a young man of thirty-six should have died without
any previous illness, caused the body of my father to be rubbed for
about two hours, which renewed its circulation and brought it to
life again. My father lived thirty-two years after that memorable
day. Without the prudence of my mother, he would either have
been dissected or buried alive. About twenty years after that
occurrence, I visited the cemetery of Père La Chaise (Paris),
accompanied by some friends. While inspecting the monuments
of some musical celebrities we heard a noise from another part of
the cemetery, whereto we proceeded without delay. When we
had arrived there we found a strong body of policemen surrounding an open grave. But in answer to our inquiring 'what had
happened,' we were simply requested to leave the cemetery at
once, which, of course, we had to do. Neither the *portier* nor any
other person connected with the burial-ground would give any satisfactory answer to our questions. We left puzzled. But a week
after, a young lady, who had been of our party the week before,
went again to the Père La Chaise, determined to penetrate the
mystery, in which endeavour she succeeded, partly through per-
suasion and partly through the gift of a twenty-franc piece to a
grave-digger, who then told her the following story :—A poor
young man of twenty-one years had been buried on the day of
our visit. When the mourners had left the cemetery the grave-
digger, who was occupied in filling up the grave, heard some noise
coming from below. He hastened to the superintendent of the

cemetery, imploring him to have the coffin opened, which, however, the superintendent could not do without the permission and the presence of the Commissaire de Police of that district. When the Commissaire appeared at last with his men, all was silent in the grave. But he had the coffin opened, nevertheless, ' to appease the mind of that poor grave-digger,' as he mockingly said. But great was the horror of the Commissaire de Police and his followers when the coffin was opened. The unfortunate young man (who was now quite dead) had been buried alive, recovered consciousness in his grave, scratched his face, bitten off the tips of his fingers, and turned around in his coffin, until suffocation put an end to his sufferings, which, if not long, must have been terrible. The Parisian newspapers did not mention the case. They were probably forbidden by the French Government to do so. But would it not have been wiser to let the whole world know of it, and thereby prevent repetitions of such dreadful occurrences? A similar case of live sepulture occurred in a village near Wiesbaden some thirty years ago, where a girl of sixteen was found with the same signs of suffocation in her coffin as those of that unfortunate young man in Paris. We are assured by a German authority that thousands of people are buried alive every year. But why should this be the case? If people must be buried before they begin to show signs of putrefaction (which seems to be the only reliable proof that life is really extinct), why not shorten their sufferings, in case of resuscitation, by opening an artery before they are buried? There is still much prejudice against the cremation of dead bodies, although two great facts are decidedly in its favour—viz., the impossibility of recovering consciousness when once inserted in the crematory oven, and the prevention of the unhealthiness which the slow process of putrefaction must entail.—Yours, etc.,

" J. H. BONAWITZ.

"London."

Professor Alexander Wilder, M.D., in his " Perils of Premature Burial," 1895, p. 16, says:—" I have often wished that the old Oriental practice of cremation was in fashion among us. There would then be at least the comfortable reflection of no liability to suffocation in a coffin. The application of fire, however, will generally rouse the cataleptic person to some manifestation of life."

Having regard to the importance of the subject, the author wrote to the hon. secretary of the Cremation Society of England, and received the following reply, dated 8 New Cavendish Street, London, W. :—

"With reference to your inquiry as to the steps adopted to prevent a person in a trance being cremated, I may say that this society has not made any special provision in that respect. You will notice, however, that before a cremation can be carried out, the cause of death must be certified without the slightest shadow of doubt by two duly qualified medical men. This being so, I think there is less likelihood of a person who is simply in a trance being cremated than buried, one doctor's certificate being sufficient in the latter case.

"(Signed)   T. C. SWINBURNE-HANHAM."

In the present state of medical knowledge on an occult subject not usually taught in the medical schools, and regarding phenomena as to which a large number of medical men are sceptical, to say the least, we fail to see how the fact of death, in the absence of putrefaction, can be certified "beyond the slightest shadow of doubt." Many of the cases cited in this volume are those regarding which the examining medical practitioners have been most sure. The Rev. John Page Hopps, in *Light*, July 4, 1896, says :—

"We are told that respect for the dead urges to burial as against cremation, but many are now very keenly feeling the

reverse of this. They can bring the mind to bear the liberation of
the body by one swift act of disintegration and purifying, but cannot overcome the shrinking from subjecting it to the foul and
lingering processes of the grave—or, perchance, to the horror of
recovering consciousness in the grave."

We take the occasion, however, to express on general
grounds our cordial adherence to the cremation movement. Mr. Hopps further states one of the strongest
arguments thus :—

" Respect for the living, too, is an urgent motive. The highest
authorities tell us that the air we breathe and the water we drink
are often contaminated by the emanations of graves. It cannot be
right that London, for instance, with all its inevitable impurities,
should add to its foulnesses that of trying to live in company with
thousands upon thousands of decaying bodies in its very midst."

To dispose of the dead decently, and at the same time
without injury to the living, is one of the first obliga-
tions of civilised communities, and cremation seems
best calculated to fulfil the conditions. Zymotic diseases, such as typhus, scarlatina, and the plague, have
been traced in certain instances to emanations from
burial-grounds.

Dr. Charles Creighton, in his "History of Epidemics in
Britain," vol. i., p. 336, says :—" The grand provocative of
plague was no obvious nuisance above ground, but the
loading of the soil, generation after generation, with an
immense quantity of cadaveric matters, which were
diffused in the pores of the ground under the feet of the
living, to rise in emanations more deadly in one season
than in another."

It would seem from these experiences as though there
was quite as much truth as poetry in Shakespeare when
he said, "Grave-yards yawn, and hell itself breathes out

contagion on the world." Before many years it is not unlikely that cremation in this as in some other countries will be made obligatory in cases of death from all infectious diseases. As the late Bishop of Manchester observed, " The earth is not for the dead, but for the living." During the thirteen years ending 1890 there were three hundred and three thousand four hundred and sixty-six deaths from cholera in Japan, and all the bodies of these persons were cremated. In India, as we have already shown, cremation is practised under most of the religious systems, as it is believed that the soul is not free from its earthly tenement until the body is reduced to ashes. The method of burning is slow and cumbersome as compared with that adopted in Europe ; but during the author's last visit to Ceylon, in the early part of the present year (1896), there was some talk of establishing a crematorium.

In " The London Burial - Grounds," by Mrs. Basil Holmes, 1896, p. 269, the question is asked :—" Are we ever to allow England to be divided like a chess-board into towns and burial-places ? What we have to consider is how to dispose of the dead without taking so much valuable space from the living. In the metropolitan area alone we have almost filled (and in some places over-filled) twenty-four new cemeteries within sixty years, with an area of above six hundred acres ; and this is as nothing compared with the huge extent of land used for interments just outside the limits of the metropolis. If the cemeteries are not to extend indefinitely they must in time be built upon, or they must be used for burial over and over again, or the ground must revert to its original state as agricultural land, or we must turn our parks and commons into cemeteries, and let our cemeteries be our only recreation grounds, which heaven forbid!"

According to Dr. Ebenezer Duncan eight thousand bodies are buried yearly in Glasgow and its neighbourhood, poisoning both air and water, and endangering the public health. The same state of things has existed in London, Manchester, Liverpool, Birmingham, and other large towns. The following resolution was unanimously adopted in the Preventive Medicine Department of a Health Congress, Glasgow, in July, 1896:—

"That in the opinion of this Congress cremation of the dead, especially in cases of infectious disease, is a natural and very desirable hygienic process, and that this Congress of the British Institute of Public Health use all proper means to urge upon the Government the desirability of their promoting a measure to enable sanitary authorities, if they so desire, to build crematoria and to conduct them under proper superintendence."

It must be allowed, however, that cremation, in spite of its obvious advantages, is not one of those movements which advance by leaps and bounds. The recent annual report of the Cremation Society of England states that during the last year there were two hundred and eight cremations in the United Kingdom—viz., one hundred and fifty at Woking, and fifty-eight at Manchester. Crematoria have recently been established at Glasgow and Liverpool.

CHAPTER XXI.

OF all the various methods that have been suggested or introduced for the prevention of premature interment, none has been attended with such satisfactory results as the erection of mortuaries (Leichenhäuser) in Germany. These structures, described in pp. 294 *et seq.*, ought to be provided, as far as practicable, in every parish, and certainly in every Sanitary District in the United Kingdom, and by the Boards of Health in the United States, and adapted to the requirements of the population. They should be of chaste and elegant design, well ventilated ; their atmosphere made antiseptic with living plants and flowers, and by plenty of light ; provided with baths and couches, and a skilled attendant—edifices where both the dead and the apparent dead can be deposited pending burial, cremation, or resuscitation. Separate compartments are necessary for cases where death has been due to accidents and for those who have succumbed to infectious diseases. Every modern appliance should be introduced for the restoration of such as may exhibit signs of returning consciousness, and of those in whom, after sufficient time had elapsed, no sign of putrefaction was observable. The temperature of the room should be kept at eighty-four degrees, as suggested by Sir Benjamin

Ward Richardson, and no interment, cremation, *post-mortem*, or embalming should be permitted until a medical examination by one or more experienced physicians showed unequivocal signs of putrefaction. Perhaps the Royal Humane Society, which during the last one hundred and fifty years has done such splendid work in restoring the drowned and asphyxiated, might be willing to extend the field of its benevolent operations to other neglected forms of suspended animation where intelligent direction and supervision is so much required.

A writer in the *British and Foreign Medico-Chirurgical Review*, 1855, vol. xv., p. 75, says :—"The earliest movements in the direction of means for the prevention of premature interments originated with Winslow in France, followed by other well-known writers upon the signs of death. It was Madame Necker, however, who embodied their suggestions in a practicable form as submitted to the National Assembly, in 1792, by Count Berchshold. In the ninth year of the first French Republic (1801) a project was entertained for the erection of six 'temples funeraires' in Paris, but came to no good, as attendant evils preponderated. To Germany belongs the credit of having executed these designs in such wise that they should not prove the positive sources of more danger to the living than could be counter-balanced by the occasional preservation of an individual from the risk of premature interment. Believing that this risk had been prodigiously diminished since the establishment of these institutions for the reception of cases where doubt of the reality of death has existed, Hufeland, in Weimar, devised the plan that Frankfort-on-the-Maine incorporated with its reform in sepulture and establishment of extra-mural cemeteries, in 1823. Hufeland's plans have subsequently been adopted and carried out in many other German States. . . . As a sanitary measure the separation of the dead from the living, especially from among the crowded poor, would be, apart from the not less important point of verification of death, an incalculable benefit. . . . It behoves us in this matter to learn another lesson from our neighbours, and to take measures to prevent the occurrence of catastrophes too fearfully horrible to contemplate in thought, too dreadful for the most vivid imagination to realise. Science can hold out no token by which to recognise the certainty of death. Sanitary police, at least in England, is indifferent about the risk of a few burials alive, and thinks it superfluous to prevent their occurrence."

That the people have a right to protection by the State against preventable sources of danger, all civilised nations have acknowledged, by the making of laws that guard their citizens from the invasion of diseases of domestic or foreign origin, as well as many other perils. But the German-speaking countries have gone further than any other in this humane direction *by recognising apparent death as a special peril to be guarded against by law*, in order to prevent living burials. For this purpose they have established mortuaries connected with cemeteries, in which the apparently dead are placed, under the observation of physicians and attendants. Here the bodies are placed upon tables, dressed in their ordinary clothes, amidst light, warmth, and ventilation, surrounded by plants and floral tributes. Thus they are kept from forty-eight to seventy-two hours, unless decomposition sets in earlier, or the death was due to an infectious disease. Further delay is allowed on application by the attending physician, or by some member of the family interested. Cords connected with an alarm bell are attached to the fingers, under the conviction that the least movement of the body would arouse the attendant in an adjoining room. No doubt these mortuaries have saved a certain number from being buried alive ; but the system can be improved by extending the observation until such time as death is certain, for experience shows that no stated limit of time can apply to all cases of trance and catalepsy, which are the chief causes of apparent death. Some of these continue for a week, and cases of even longer duration are not unknown. It often happens that returning vital activity consists merely in scarcely perceptible movements of the eyelids or the mouth, a change of the complexion, slight moisture on the face, or a faint action of the heart, or a warmth in that region, or feeble thoracic movements—all of which might escape observation until the allotted time had expired, and no contrivance, however delicately adjusted, could announce their presence. Time alone will test the existence of life or death in such cases.

The extensive literature on this subject shows that the struggle to bring about the existing mortuary system in Germany was kept up for many years before it obtained its measure of success. It was legalised about the year 1795, after the physicians of Germany, France, and Austria had shown the absolute necessity for it.

Mortuaries have continued in high favour with the people wherever they have once been properly established ; none, so far as the author has been able to learn, have ever been abolished. At the present time the city of Munich is constructing a mortuary at the Southern cemetery upon a costly scale, surpassing in sumptuous accessories anything of the kind before attempted in Germany. It will be not unworthy of the public buildings of the city. This is an emphatic endorsement of the necessity of the system by a people that for more than fifty years has given it a thorough trial ; and it is a strong argument for its adoption elsewhere.

The question suggests itself here : Why should not the English-speaking peoples accept the long experience of a philosophical, painstaking, clear-minded people like the Germans, supported as it is by many sanitary and medical authorities in France, England, and the United States, and establish these institutions in connection with existing cemeteries, with such modifications as national habits, local tastes, and customs may dictate ?

The following practical suggestions are from a paper in the *Medical Times*, vol. xvi., No. 415, p. 574, September 11, 1847, entitled, "On the construction of houses for the reception of the dead ; and on the means to be used for the recovery of those who are only in trances or fits, or in whom life is only impassive," by Robert Brandon, Esq., Great Russell Street, Bloomsbury :—

"*DUBIÆ VITÆ REFUGIUM;* OR, ASYLUM FOR DOUBTFUL LIFE.

" The building should be large enough to provide means for resuscitation, and have room enough for the deposition of bodies when epidemics are prevalent. There should be hot baths, for these often are alone enough to recall the vital spark ; and a kitchen to prepare nourishment for those who are recovered, and for the porter and other officers who would live on the building. The room for the deposit of the bodies should communicate with the porter's room by means of a glass door, and every body should have a wire fixed to the feet and hands, in communication with a bell, which bell must ring in the porter's room, in order to warn him should there be any motion in those thought to be dead. There should be men and women on the premises to use friction, a galvanic machine, and the implements necessary for transfusion and artificial respiration. As the usual and accepted signs of death are not signs to be relied on, so is decomposition a true sign, and none should be buried until this be present ; but as the presence of decomposed animal matter would be injurious, not only to the inmates of houses, but to the surrounding inhabitants, and as it is inconvenient to the poor man who has but one room to keep a body in that room, where he and his family eat, drink, and sleep, asylums for the reception of those thought to be dead should be constructed, and are absolutely necessary. Nor is it enough to wait for decomposition, but we should endeavour to prevent this by endeavouring to restore vitality by means of hot baths, external heat, artificial respiration, galvanism, or transfusion ; the first of these is oftentimes enough. Now, I think it probable that many persons would be recovered, thought to be dead, for, out of a number of those reputed dead, a certain number have recovered—some by the sticking of the pins into them which fixed the shrouds, some under the surgeon's knife, some from delays in the burial, and others from the accidental over-turning of the coffins, as we learn from a paper published on premature burials. Some time since a woman was kept above ground for a considerable time, as medical men could not decide if she were dead or no. And at Constantinople a sailor the other day was attacked with apoplexy, and a vein was opened in his arm ; no blood came, and the man was thought to be dead, but on the road to the grave blood began to flow, and the sup-posed dead man recovered. There is now living in Brussels a man who escaped from the grave; and another built a house at Cologne to commemorate his escape. These cases will be enough to show that we have no certain sign of death but decomposition ; and, if this be true, we must have asylums for the reception of bodies previous to decomposition, and for the application of means which can do no harm, and may do much good, such as those before indicated. Medical men think that the absence of respiration and want of heart's action, with loss of motion and sensation, are signs of death ; but this is not the case, for many bodies which have been drowned have all these signs present and yet recover. Again, infants are often born without any action of the heart or lungs, and yet are recovered by very simple means, such as the hot bath ; and I myself have recovered persons by stimulants who were thought to be dead. Many may be recovered by transfusion (first introduced into this country by the celebrated Dr. Blundel) when the heart still palpitates, but the brain is insensible ; or by stimulants given at that period ; or by hot bath, and the external application of heat ; by galvanism, where other means have failed;

and these can do no harm. Since the brain is insensible, there can be no suffering; and many lives will be saved by perseverance, and the skilful application of means which have succeeded in isolated cases. Buildings for the reception of those thought to be dead should be placed in cemeteries.

"I divide life into active and passive. Life is active when man is in the enjoyment of all his faculties, intellectual and moral; when the various organs necessary for circulation and respiration are in play; when there is sensation, perception, and motion; and when the sphincters are not relaxed. Passive life is that state hitherto called death; but, according to me, death is decomposition.

"Nor should we despair at any period previous to this, since we can give motion by galvanism; blood by transfusion; respiration by artificial respiration; heat by this and the external application of caloric; and by stimulants we can keep up that action which has been excited by other means. Nor must we despair if we do not at once succeed in our endeavours to recall life, for perseverance often accomplishes that which at first sight seems impossible.

"Men have recovered from simulated death after being in the sea twenty minutes, and I see no reason why, after disease, men may not also be recovered from a state resembling death. Many who are left as dead are only in fainting fits, some are in trances; and graves have been opened where the buried man has been found to have eaten portions of his own flesh, which of course he could not have done unless recovery had taken place. How horrible to think that we may awake up in our graves tormented with the pangs of hunger, unable scarce to breathe, and finding all escape from our narrow cell impossible; the prisoner in his grave has nought to do but to commend his soul afresh to his Maker, and lay himself down to die! May not much of this be prevented by asylums for doubtful life, by the application of reagents, and by building vaults in our cemeteries instead of graves? I earnestly hope that the day has arrived when we see these things in the proper light; when our church-yards will be no longer overloaded with the remains of those who, perhaps, might have lived had they been left a little longer above ground—had they been transfused, or even buried in vaults instead of graves, with a guardian to watch over their mortal remains! Life may exist, but not be evident; but the non-evidence of life is no proof of death, as many have been recovered in whom life was only latent—in whom there was no action of the heart, no respiration, no motion, no sensation. This has happened after drowning, in infants born asphyxiated, in women after flooding, and would happen much more often were the proper means applied in all cases to recall life, and to ascertain those who may be recoverable. Simple inspection is not enough to decide if a man be dead or not, because persons are often only in trances or fainting fits when they are thought to be dead; and I wish to insist on the fact that there is no sign of death but decomposition, and that, therefore, none should be buried until this sign be present, nor until an attestation of the presence of decomposition be given by some surgeon."

Referring to the universal fear of burying relatives alive, the *Lancet*, September 20, 1845, vol. ii., p. 321,

20

observed :—" It is but little use to descant upon an evil
without pointing out a remedy.   In Frankfort, Munich,
and in various other towns, houses, properly situated,
have been fitted up for the temporary reception of the
dead.   Corpses are there deposited immediately after
death, and taken care of until the signs of decomposition
have become unequivocal, medical assistance being at
hand should symptoms of vitality manifest themselves.
By this simple plan all the objections which attend on
the retention of the dead in the dwellings of the poor
may be obviated, and at the same time their dread of
burying their relatives whilst still alive respected.   This
plan is evidently much preferable to that which is
followed in France.   In the latter country, in the large
towns, there is in every district a medical inspector of
the dead.   The inspector is informed of the death as
soon as it has taken place, and within a very limited
time is bound to inspect the body and give a formal
certificate.   This guarantee having been obtained, the
inhumation of the deceased is enforced by law within
two or three days of the death.   Notwithstanding this
precaution, cases have occurred, even during the last
few years, which appear to prove that inhumation has
taken place before life was quite extinct.   We doubt,
also, whether such early interment could under any
circumstances be enforced in our own country.   Some
modification of the German plan is evidently what we
must look for in any system of legislation which may
hereafter be decided on."   These admirable suggestions
from the leading medical journal were made more than
half a century ago ; since that time, every year has
brought to light cases of living burial, and confirmed

the urgent need of reform ; but nothing has been done until quite recently to awaken public attention to their importance. The subject is of such a gruesome, unpleasant, and depressing character that few people care to have their names associated with a movement of this character, beneficent though it is, and certain to save thousands of unfortunate people, particularly women and children (who are more especially liable to various forms of suspended animation), from such tragic occurrences.

The *Undertakers' and Funeral Directors' Journal*, August 22, 1895, referring to the fact that in 1892 thirty - one thousand eight hundred and ninety - two inquests were held in England, and to the urgent necessity for the erection of mortuaries, says :—

"The bountiful, or private enterprise, should provide these mortuaries. But once let their necessity be recognised and the scheme approved,—fashion leading the way,—then undertakers would readily supply what was wanted. If not, then the local authority should take the initiative. Mortuaries are sadly needed almost everywhere for present purposes, as newspapers constantly affirm. In providing them, care should be taken to build with an eye to future requirements when it shall become customary if not compulsory to remove the dead from among the living within a reasonable time after death.

"It is merciful sometimes to be inexorable, and what a lot of willing and unnecessary discomfort and risk would be saved were it possible and the practice to find a temporary resting-place for our departed friends till we are ready to carry them befittingly to the tomb."

MORTUARIES OF LONDON.

Each of the sanitary districts in the Metropolis is supposed to have a mortuary of some kind for the reception of bodies from hospitals, infirmaries, hotels, private houses, as well as from the river and streets, or in transit to and from foreign countries, where they are kept without charge for about five days, unless the public health requires earlier interment. Hospitals, hotels, and families are thus relieved of the presence of corpses, for convenience, and for purposes of inquest. The mortuaries are nearly all plain, gloomy, and depressing structures of brick. The best of them comprise a coroner's court-room, coroner's private room, the caretaker's rooms, waiting room, *post-mortem* room, chapel, and viewing room connected. There is no physician in attendance, and no autopsies are performed except by surgeons upon their own cases, or for purposes of inquests. There are no appliances or conveniences for resuscitation, as all the bodies are regarded as dead, having been, for the most part, certified as such by a medical practitioner, the exceptions being such as are taken from the water or street by the police, or left there for inquest. The buildings are usually well lighted, and some of the rooms contain fire-places, but they are devoid of taste or ornamentation of any kind. The bodies are kept in coffins, which, if there is any odour proceeding from them, are screwed down. Permission is afforded for inspection by doctors or by any of the family of the deceased on application to the keeper. These mortuaries are kept clean, and decent and respectful treatment of the bodies is enforced by regulations.

The London County Council issued a return (No. 157) dated March 9, 1894, in pursuance of the Public Health (London) Act, 1891, relating to coroners' courts, mortuaries, etc., from which it appears that there were fifty-one mortuaries in the sanitary districts of London up to September 30, 1893. In most of these the accommodation is described as "sufficient," "good," "well arranged," "excellent," "convenient." Others are of an opposite character. The one attached to the Town Hall, Holborn district, is reported as "very small (about nine feet by nine feet), inconvenient, and badly situated." In the Poplar district the mortuary "is an old crypt, quite unfit for the purpose, and has no convenience for *post-mortems.*" At Ratcliffe, in the Limehouse district, the mortuary "consists of a railway arch, and is very unsuitable." "There is a very small mortuary in the church-yard" at Shadwell. The mortuary under the church-yard of St. Martin's Church (St. Martin's-in-the-Fields) is reported "very imperfect." The one in the Southern Coroner's district is situated under a railway arch, and there is no mortuary-keeper. At St. Paul's, Deptford, the mortuary contains only one room, which serves for mortuary and *post-mortem* room. Plumstead is possessed of an underground mortuary in the church-yard, reported as "unsatisfactory." The Lewisham district has an "unsuitable" mortuary at the cemetery. Rotherhithe has "an inadequate mortuary in the old burial-ground." At St. George the Martyr (Southwark) the mortuary is reported to be "inadequate and unsuitable." In the Strand district there is "no proper mortuary, but a small dead-house attached to the Savoy Chapel is used." Eltham, Lea, and Kidbrooke, in the Plumstead district, have no mortuaries. The part of Lambeth, S. and S.E., up the Clapham and Kennington Park roads, is without a mortuary, and *bodies awaiting inquest are kept in private houses.* Nor are there any mortuaries in the Greenwich district (Hatcham), Wapping, or Mile End Old Town. Arrange-

ments are reported to be in progress for the enlargement
of some of these establishments and the erection of
others.

No resuscitations are reported from any of these places,
except in the case of Ernest Wicks, a boy two years old,
who was found lying on the grass in Regent's Park
apparently dead, and resuscitated in St. Marylebone
Mortuary (after being laid out on a slab as dead) in
September, 1895, by the keeper, Mr. Ellis, assisted by
Mrs. Ellis. When the doctor arrived, the child was
breathing freely, though still insensible. The child was
taken to the Middlesex Hospital, and was reported by
the surgeon to be recovering from a fit.

The London mortuaries stand well in the estimation
of the authorities, medical practitioners, and the people,
on account of their usefulness and convenience in re-
lieving hotels and private houses of the dead pending
funerals, and in cases of deaths from infectious diseases,
as well as from accidents and acts of violence (amongst
which suicides are included) which require investigation.
In consequence of this, there is a disposition on the part
of the authorities to enlarge and improve the older and
smaller ones, and to introduce the later conveniences.
Those in St. Marylebone and St. Luke's are the latest
examples, and could, with comparatively little outlay,
be rendered creditable and useful establishments.
First of all, they require the means of resuscitation,
such as are in use at the Royal Humane Societies'
Depôts, and at the German mortuaries; also baths,
couches, plants, flowers, and mural ornaments, with
a skilled nurse or caretaker, and a medical practitioner
either on the establishment or within telephone call.

A fundamental regulation should be added to the standing orders that, when there is no sign of decomposition, bodies should be treated not as dead. but as sick needing attention, and to be kept under careful observation. Such simple and inexpensive alterations, gradually introduced by County, Parish, and District Councils, would, in the course of time, bring about a greater respect for the dead, with proper consideration for the apparently dead, besides increasing the feeling of the sanctity of human life. In the course of time these improvements would educate the public, and lead to the erection of new and handsome structures of beautiful design, with appropriate artistic decorations, such as are to be found in Munich and other parts of Germany.

The *Medical Times*, September 5, 1896, p. 569, says :—

"In a recent issue of the *Nursing Record*, there is an interesting article on hospital mortuaries by a special commissioner. . . . At Guy's the mortuary only contains room for one body. There is a bier, covered by a cradle and a red and white washing pall, and over this is a shelf, on which are placed a cross, fresh flowers, and candles. At St. Bartholomew's the mortuary itself is certainly not a place where one would care to find one's dead. The bare, whitewashed walls, the sloping floor, the black lidless shells, covered by white sheets, would depress most people even if they had no special interest in them. That this is felt to some extent by the hospital authorities is evident from the fact that, when a member of the staff dies, they do their best to make other arrangements for the disposal of the body until it is removed from the hospital. There is an hospital not named [continues the *Medical Times*] where the only place available as a mortuary is the wash-house. It would appear that the managers of metropolitan hospitals do not believe in the reality of death-counterfeits, and therefore make no arrangements for resuscitation."

MORTUARIES IN THE PROVINCES.

With the object of ascertaining the utility of these establishments, the author wrote to the clerks or other officials in all the larger towns in the United Kingdom, fifty in number, requesting copies of the regulations, reports, etc. To these communications twenty-four replies were received. Of these, only three sent copies of reports, furnishing particulars of the number of bodies received, and the number of inquests and *post-mortems;* three sent copies of regulations; and the remainder do not publish either reports or regulations. One, however (Poplar), states that the by-laws in use are approved by the Local Government Board. The Chief Constable at the Town Hall, Salford, writes, July 26, 1896—" There are three mortuaries in the borough, but a separate record of the bodies laid in the mortuaries is not kept, and no papers exist respecting them." Mr. Hagger, the Vestry Clerk of the Parish of Liverpool, says—" I know of no public mortuary in Liverpool which is considered to be of such importance as to call for anything in the shape of periodical reports." Mr. R. Davidson, Governor of the City Parish Poorhouse, Glasgow, writes, July 27, 1896—" I have never had any reports relating to the mortuary here." Mr. J. Jackson, Chief Constable, Sheffield, writes, July 29—" We have never had papers or reports connected with it (the mortuary), except the ordinary rules and regulations for preserving decency, cleanliness, etc." Similar replies were received from Manchester, Swansea, Scarborough, Wigan, Bristol, St. Mary's (Islington), Dundee, and Catford. Mr. Robert Clinton, Master of the Bethnal

Green Workhouse, writes, July 30—"That their mortuary has not been the subject of any reports," and continues, "The subject of persons being buried alive is a very important one, and should arouse the interest of every intelligent person. Some method ought certainly to be devised that will prevent anyone being subjected to so horrible a fate."

IRELAND.

The following extracts are from the report by Dr. J. E. Kenny, M.P., Coroner for the City of Dublin, received in January, 1894 :—

"There are no local laws in Dublin or in Ireland relative to the mode of disposal of the dead, but the Sanitary Acts, which refer to the United Kingdom of Great Britain and Ireland, can be availed of when necessary to compel the burial of the dead within a reasonable period, on the ground that an unburied body is a nuisance dangerous to public health.   There is, however, no fixed period.   Among Roman Catholics it is customary to bury the dead on the third or fourth day after death, but there is no hard-and-fast rule. . . .   The local burial authorities usually require a medical certificate of death before opening the grave, but there is no legal sanction for this, and it is merely the custom.   The coroner's order for burial where an inquest is held does away with the necessity of such certificates as those above referred to, but *post-mortem* examinations in these cases are the exception, not the rule.   A good many, however, are held on those who die in local hospitals, when the consent of the relatives or friends can be obtained. I have not heard of any case of cremation in Ireland, and earth-burial is the universal practice. Occasionally, when so ordered by the will of the deceased, a body is removed to England for cremation. I am myself rather in favour of cremation as a more scientific and safer method of disposing of the dead.

"There are no chambers (mortuaries) of the kind referred to in this question in Dublin, nor, so far as I know, in Ireland. I know of no law as to the signs of death which must be recognised to exist before burial is permitted, nor is there any officer on whom is thrown the duty of ascertaining or deciding whether such exist or not.

"If cremation be generally adopted, it ought not to be performed earlier than the third day after death, or perhaps not until some unmistakable sign of decomposition has set in. I think this rule of some such sign of decomposition setting in ought to apply to all methods of disposal of the dead. Whenever well-marked warmth of the body exists after apparent death, burial of any kind ought not to take place until after a full and exhaustive examination by a competent authority. In all doubtful cases I would suggest the application of either a hot iron to some sensitive part of the body, or that a small incision should be made over the course of some small artery, a person being left to watch the result for some time in the latter case, so as to take proper precautions against hæmorrhage, should the person be not really dead. It might perhaps with advantage be made the law that in every case of death or supposed death the body should be viewed by a medical man, who, having satisfied himself that death had taken place, would sign a certificate to that effect. If I understand rightly, such is the law in France. I would, however, be opposed to any law making an autopsy necessary in every case. The existence of such a public officer as a coroner is undoubtedly of advantage in reference to cases of sudden death or supposed death, as it is among such cases that mistakes are most likely to occur. I can see no objection to the establishment, at the public expense, of chambers for the reception of dead bodies under certain circumstances."

In reply to a similar inquiry Sir Charles A. Cameron, Superintendent Medical Officer of Health, writes, August 10, 1896—"There is no public mortuary in Dublin, but we are taking steps for the establishment of one."

It need hardly be said that the mortuaries described in these reports have little in common with certain *Leichen-häuser* of Germany or the *Mortuaires d'Attente* urgently called for by various writers of France, and proposed to be erected. The English mortuaries may more appropriately be described as *morgues* or depositories for the homeless and neglected dead—useful for this purpose, but in no respect fulfilling the requirements of the present day. Without skilful attendants and scientific appliances for the restoration of suspended life, to which all are liable, the apparently dead, if deposited in such chilling establishments, would, through neglect, be more likely to lose what spark of life remained than to have it kindled into a flame and recover. The erection of mortuaries for the sake of death-counterfeits, and in order to give peace of mind to doubting friends, would no doubt be opposed

chiefly on the ground of expense. The outlay must
come from the pockets of the rate-payers, who have been
accustomed to accept the cursory inspection of "the
corpse" and the certificate of the doctor as a satisfactory
solution of any misgivings as to the actuality of death.
Under the circumstances it would not be surprising if the
unreflecting majority preferred to take what they would
consider to be an infinitesimal risk rather than to incur
the expense of the necessary outlay. This volume has
been written to remove such apathy, and, if possible, to
arouse public attention to the subject; and if the facts are,
as the author believes, absolutely true, and the danger
real, other and abler contributions furnishing the results
of wider and more extensive investigations may be
expected to follow. It is believed that the expense of
constructing tastefully designed mortuaries in all populous districts could be met by a rate of from a farthing to
a penny in the pound, and in the smaller or thinly
populated districts groups of parishes could unite in
providing such useful institutions. At present, under
existing customs, probably ten times the amount required
is annually expended in funeral trappings, mourning
habiliments, costly wreaths, and ornamental monuments (mainly for the purpose of ostentatious display)
than would provide temporary resting-places for the
real and apparently dead in every part of the United
Kingdom. The erection of such establishments, where
the fact of death in every case could be unequivocally
demonstrated before burial or cremation, would remove
an ever present and consuming load of anxiety from
the hearts of thousands of sensitive souls.

CONTINENTAL MORTUARIES.

The author is indebted to a "Treatise on Public Health," by Albert Palemberg and A. Newsholme, London, 1893, for the following details :—

BRUSSELS.

" This city possesses two mortuaries to which bodies are conveyed from confined houses. One of these, within the town, only receives the bodies of persons not having died of an infectious disease ; all others are conveyed to the mortuary at the Evère Cemetery. . . .

" In times of epidemic the removal of corpses to the mortuary is compulsory, and so also in other cases where the medical health officer decides that it is necessary. No corpse, without special permission, can be kept in the mortuary more than forty-eight hours after death, but this interval can be shortened or lengthened by special order."

PARIS.

" By a decision of July 21, 1890, the Municipal Council of Paris has decided to establish a mortuary in each of the cemeteries of the east (Père La Chaise) and the north (Montmartre). . . . The mortuaries are not available for the bodies of persons having died from infectious disease.

" Bodies are only admitted to the mortuary—(1) On the written application of the head of the family or some other persons competent to undertake the funeral. (2) On the production of a certificate of death from the doctor who attended the patient, stating that the death was not caused by infectious disease.

" Up to the present time (1893) these mortuaries do not appear to have been of great service, owing to the unwillingness of families to part with their dead before the time of interment.

" ' La Morgue.'—This establishment only receives bodies on which a *post-mortem* examination is required, and the bodies of unknown persons, placed there for recognition. In the hall where the bodies are exposed, the temperature is kept several degrees below zero by a system of refrigeration, thus retarding putrefaction.

This system would, in consequence of the low temperature, greatly retard or prevent the revival of persons who may only be in a state of torpidity from submergence, or of trance or catalepsy, who could be resuscitated if warmth and other proper means were promptly applied to them."

BERLIN.

" In some of the cemeteries mortuaries have been built, which are placed at the disposal of the public by the authorities, with the understanding that the corpses shall be taken from them as soon as possible.

" The bodies of the poor are first placed in the depository of the old cemeteries, within the city enclosure, whence they are removed by night in carriages kept for the purpose to the mortuary in the large cemetery outside the city, to be buried the next day. The Jews have built a mortuary chapel in their new cemetery at Weissensee, which fulfils all the conditions required by modern hygiene, and contains everything necessary for washing, isolating, and enveloping the bodies.

" A new establishment, which answers its purpose perfectly, has been built in the old cemetery—Charité—and is used for inquests, *post-mortem* examinations, etc., also for the exhibition of bodies of unknown persons. The bodies are preserved from putrefaction by an apparatus in which refrigeration is produced by ammonia and chloride of calcium, as the Morgue in Paris."

VIENNA.

" There is a mortuary in each district of the city to which are brought corpses belonging to families who have imperfect accommodations.

" The district doctor must decide whether removal is necessary, as it is his duty to register deaths and their causes. He should at the same time examine into the state of the dwelling from a sanitary standpoint.

" In cases of sudden death, and when the cause of death is not apparent, a *post-mortem* examination must be made.

" The bodies of persons who have died from infectious disease must not be taken to the common mortuaries, but to one built in the common cemetery.

"Bodies must not be buried in the city. The principal cemetery is at Kaiser-Ebersdorf, north-west of the city, and cost four millions of marks."

STOCKHOLM.

"Every parish possesses a mortuary vault. According to the regulations of the Health Commission, bodies must not remain there more than forty-eight hours in the hot season, and seventy-two in the cold weather."

The first modern mortuary was opened at Weimar, Germany, in 1791.

In a "Handbook for Travellers in Europe" for 1890, by W. Pembroke Fetridge, p. 622, is the following description of the model mortuary in Weimar :—

"The New Church-yard is a sweet place of its kind. Here may be seen an admirable arrangement to prevent premature burials in cases of suspended animation. In a dark chamber, lighted with a small lamp, the body lies in a coffin. In its fingers are placed strings, which communicate with an alarm clock ; the least pulsation of the corpse will ring the bell in an adjoining chamber, where a person is placed to watch, when a medical attendant is at once supplied. There have been several cases where persons supposed to be dead were thus saved from premature burial."

The *Middlesborough Gazette* of 11th October, 1895, says :—

"Those who have visited burying grounds in some parts of the South of England are well aware that tombs made in the shape of ' waiting rooms ' are largely in vogue with the well-to-do classes. One in a little church-yard in Sussex was elegantly fitted up. The coffins were placed on one side of the well-lighted vault, while on the opposite side was a couch, chairs, and a table, together with books. The relatives of the deceased—eccentric they may have been, we are not prepared to say—visited the vault, access to which was gained by a flight of steps, and there passed much of their time in reading, the ladies doing needle work. But this sort of thing is only for the rich. The poor must be protected from being buried alive by other and more economical methods—namely, by stricter attention to the actual and unmistakable evidences of death, and by careful registration on medical certificates only."

It would appear by the following announcement, that an effort is being made to supply one of the several properly fitted mortuaries needed in the French capital :—

"The *Pall Mall Gazette* of September 21, 1895, announces a decided novelty in the way of limited liability companies—the Mortuary Waiting-room Company, which, it says, is on the point of being floated in the French capital. Our contemporary says that the amount for subscription is stated to be £20,000, and dividends at the rate of at least 100 per cent. may, it is claimed, be confidently looked for. The company undertake to provide separate waiting-rooms, of two classes, in a large mortuary building. The alleged corpse will be comfortably deposited there upon a couch, and carefully looked after till the fact that it is a corpse shall have been established beyond question. The waiting-rooms will be tastefully decorated, with everything about them to welcome the revived tenant agreeably back to life. It is interesting to hear that no shareholder's heirs will be allowed to visit him."

Some sanitarians and funeral reformers urge with much reason that the presence of the dead should not be allowed to endanger the health of the living, and recommend that, if death has occurred from infectious disease, the body should be covered with charcoal and conveyed at once to a mortuary chamber ; and others advise early burials for all as soon as possible. If, however, this volume has not demonstrated the danger of such early burials, except where decay of the earthly vesture is visible, it will have been written in vain.

The following recommendation from a well-known physician and surgeon appears in *London*, p. 613, September 27, 1894 :—

"Coroners' Courts and Mortuaries," a paper read at the Hygiene Congress at Buda Pesth, by W. J. Collins, M.D., M.S., B.Sc., D.P.H. (Lond.), L.C.C.

"I therefore hold that every inducement should be held out to the poor by local authorities, by the provision of decent, suitable, and attractive mortuaries, to allow their dead to be removed from danger to the living to a place where sentiment shall be respected and sanitation satisfied."

THE UTILITY OF MORTUARIES.

During the discussion on Premature Burials in the press, the erection of mortuaries (chambres mortuaires d'attente) has been objected to (1) on the ground of expense to the ratepayers ; and (2) because the results by way of resuscitation of those constructed in Germany have not justified the cost of their erection and maintenance, and that if they had not already been in existence they would not now, it is said, be established. The most recent investigations on this subject have been made by Monsieur B. Gaubert, the results of which appear in his work, "Les Chambres Mortuaires d'Attente," a volume of 308 pages, published in Paris, 1895. The author shows by the citation of facts that both in France and Germany numerous cases of resuscitation of persons certified as dead, and deposited in mortuaries, in spite of many drawbacks connected with their management, have occurred, and that their continuance is amply justified on the ground of utility. In the report of the Municipal Council of Paris for 1880, No. 174, p. 84, is a letter from Herr Ehrhart, Mayor of Munich, May 2, 1880, who says :—"The lengthy period during which these establishments have been utilised, the order which has always prevailed, the manner in which

21

the remains are disposed and adorned, *the resuscitation of some who were believed to be dead*, have all contributed to remove any sentimental objections to these establishments. The bodies are transported to the Leichenhäuser twelve hours after death, without the least opposition on the part of the relatives." The expense of these institutions would, no doubt, in the aggregate be a considerable sum, but not nearly so large as that voted for the erection and maintenance of public libraries, now so common ; but in the presence of so serious and real a danger as that of living burial, to which any of us is liable, it is hardly worth considering. For peace of mind the cost of such insurance would be cheerfully paid by thousands, and ought to be provided for the poor and for those who would in time come to value it. This is a matter that might appropriately be taken up by the County, District, and Parish Councils and Boards of Guardians, under the powers granted to them by the Local Government Act of 1894.

Dr. Josat, in his treatise " De la mort et de ses caractères," shows by numerous arguments and examples that, as there is an interval or condition provided by nature between disease and health known as *convalescence*, and the transition between the one and the other is preceded by a variety of phenomena known as a *crisis*, so there is an interval between the termination of a fatal malady and real death (erroneously described as the agony), the symptoms which denote intermediate or apparent death. But while the result of an error may be of little moment in the first case, it may in the other become disastrous, by abandoning the dying before absolute death. It is during this interval, between (so called) death agony and absolute death, which sometimes has been known to last a week, that the transfer to a suitable mortuary should be made.

The following may be cited as typical illustrations of the utility of mortuaries in discovering the existence of life after apparent death.

H. L. Kerthomas in "Dernières Considérations sur les Inhumations Précipitées," Lille, 1852, p. 17, relates that—

"At a hospital in Liege two house-surgeons were at the 'Salles des décades' in pursuance of their anatomical studies, when, hearing at one side of them a noise like stifled breathing, great was their fear! Still they coolly finished their examination, and then discovered the supposed corpse moving convulsively amongst his dead companions ; but, thanks to efficient help, he was completely restored to health." (The above occurred in 1847.)

M. B. Gaubert, in "Les Chambres Mortuaires d'Attente," records the six following cases:—"On the 25th of January, 1849, the *Journal des Débats* recorded a fact somewhat similar to that which lately disturbed the town of Perigueux :—

"'MUNICH.

"'A young man who was asphyxiated by charcoal had been declared dead by the doctor. After they had been watching the body twenty-four hours at the mortuary chamber, the family caused it to be carried to the church, where it passed the night without the customary caretaker. The next morning "the corpse" was found bathed in its own blood, and the floor of the church was stained. Restored to consciousness during the night and not having any help, the poor young man had succumbed to hæmorrhage, brought on by the incisions which they blindly practised on the body of the supposed dead one to make sure of his death.'

"'The mother of a family had just lost her child, aged five years. She carried to the Leichenhäuser a heart broken by grief, cherishing the vague hope in the depth of her love that this separation would not be the last. According to habit the families of Munich exposed the corpse in a mortuary chamber amidst flowers and trees, and surrounded by a circle of light. The Leichenhäus then appeared to have lost its habitual funereal character—for it had quite a festive air. The poor mother passed the night amidst tears and prayers, waiting with anxiety and hoping for the arrival of the good news. The next morning a workman of the Leichen-häuser knocked at the door of the house with a large bundle which he carried in his arms ; a few seconds after, the mother pressed to her heart the resuscitated child which she was told she had just lost. The transports of joy she experienced were so great that she fell down dead. The child had come to life in the mortuary by himself, and when the keeper saw it, it was playing with the white roses which had been placed on its shroud.' (P. 179.)

"The same recent writer quotes the following on the testimony of the surgeons Louis and Junker :—

"'SALTPÉTRIÈRE.

"'A young country girl,' said Surgeon Louis, 'strong and vigorous, twenty-five years old, left on foot from the Hotel Dieu, Paris, where she had been resting the night before, and came to Saltpétrière. The fatigue of the journey induced an attack of syncope on her arrival. They put her on the bed, and with cordials and warmth she revived, but at the end of an hour she had another attack. They thought she was dead, and carried her to the mortuary. After leaving the body—it had remained there some time—they carried it to the amphitheatre. The next morn-ing a young surgeon said he had heard plaintive cries in the amphitheatre, and his fear had prevented him from coming to tell me. I went into the amphitheatre, and saw with sorrow that the poor girl, who had vainly struggled to free herself from the sheet which enveloped her, was now quite dead. She had one leg on the floor, and an arm on the seat of the trestle of a dissect-ing table. I here recall the feelings of horror with which I was agitated on this occasion. I doubt if there ever was a sadder or more touching spectacle than this.' (P. 187.)

"BERLIN.

"'A Berlin apothecary wrote to me lately'(says Dr. Lénormand)'in this town to the effect that during an interval of two years and a-half, ten people stated to be dead had been recalled to life. I shall quote only the following :—

"'SOLDIER OF THE GUARD.

"'In the middle of the night the bell of the vestibule rang violently. The caretaker, who had only entered on duties within a few days, much startled, ran towards the mortuary. As soon as he opened the door he found himself confronted with one of "the corpses" enveloped in his shroud who had quitted his bier and was making his way out. He was a soldier of the guard believed to be dead, and he was able to join his regiment five days later.' (*Ibid.*, p. 180.)

"FRANKFORT-ON-THE-MAINE.

"Dr. Josat said that during his sojourn in Germany, Herr Schmill, director of the mortuary at Frankfort, related to him a case of apparent death which occurred under his own eyes.

"'In the year 1840, a girl of nineteen years died of acute pleuro-pneumonia. Her body, during very hot weather, was exposed in the mortuary for a period of eight days in a state of perfect preservation. Her face retained its colour, the limbs were supple, and the substance of the cornea transparent, whereas in ordinary cases decomposition shows itself on the third day. The parents could not reconcile themselves to have their daughter buried, and found themselves much troubled. Finally on the ninth day the supposed dead suddenly awoke without any premonitory indications of life.' (*Ibid.*, p. 180.)

"BELGIUM.

"There was a case at Brussels in January, 1867, of a person who returned to life just as the bearers arrived at the mortuary.

"'A workman of the suburbs, employed by a firm of carriers, fell ill, and in a few days died. This suddenness of the death caused doubts as to its reality, and after the usual delay he was taken to the mortuary connected with the cemetery. The body was left for a few days' observation. As soon as they arrived a noise escaped from the coffin, and arrested the attention of the people present. At once they hastened towards the coffin, and tried to restore him, and in a short time he came to life. The same evening he was able to return to his home. On the following day he went himself to the authorities to annul the record of his supposed death.'" (P. 182.)

M. Gaubert continues :—"We have collected in Germany fourteen cases of apparent death followed by return to life in mortuaries, in spite of all that has been done for the prevention of such occurrences." (P. 182.)

CASSEL.

" Dr. E. Bouchut, in ' Signes de la Mort,' 3rd edition, p. 50, relates that an apothecary's assistant had an attack of syncope, which continued for eight days, when he was apparently dead, and was removed to the mortuary of the Military Hospital, Cassel, where he was covered with a coarse wrapper and left amongst the dead. The following night he awoke from his lethargy, and, on recognising the horrible place where he was, dragged himself to the door and kicked against it. The noise was heard by the sentinel, aid arrived. and the patient was put in a warm bed, where he recovered. Dr. Bouchut says that, if he had been swathed in tight bandages, his efforts at release would have been futile, and he would have been buried alive."

LILLE.

The Paris *Figaro*, March 31, 1894, on the authority of the *Progrès du Nord*, April 2, 1894, reports that :—

" M. Vangiesen, aged eighty-one years, awakened from supposed death on the flagstones of the mortuary at the Charité Hospital at Lille."

The *Undertakers' Review*, January 22, 1894, reports that Lena Fellows, aged twenty-two years, a servant in the employ of A. R. Knox, of Buffalo, fell dead, as was thought, while at work on December 8. The remains were taken to the morgue in a coffin, but next morning when morgue-keeper McShane began to lift the supposed corpse into the refrigerator he found that the woman was alive. It was a case of catalepsy.

The case of a child found apparently dead in Regents' Park, London, and carried to the Marylebone Mortuary, where it subsequently revived, has already been noticed. The incident caused a good deal of comment, and suggested, doubtless, to the reflective reader that other cases of suspended animation might have a less fortunate issue.

It is quite impossible on the Continent for an enquirer, as the author knows from experience, to obtain reliable information with regard to what takes place within the walls of mortuaries, because of the numerous officials and others who are interested in covering up any errors of previous death-certification that may come to light in them. These comprise the health authorities, and the police in places where the latter regulate funerals, as well as the physicians, whose credit is at stake, and the nurses and undertakers. In many districts in Germany the original object of the mortuaries—the prevention of premature burial—advocated by Hufeland and others, has not been kept in view, but the edifices have rather been used for the convenience of the undertakers and their assistants, the bodies in many cases being removed before actual dissolution was established by evidence of putrefaction. This will need to be guarded against by more careful supervision.

CHAPTER XXII.

CONCLUSION.

IT is universally admitted that nothing is less certain than life ; and if the reader will weigh the facts, which it has been the authors' intention to understate rather than overstate, he will rightly conclude that nothing is more uncertain than the signs which are ordinarily accepted as indicating death. It would have been easy to fill a much larger volume than this with reports of authentic cases of premature burial, and narrow escapes from such terrible mischances, and with more detailed results of the authors' researches on the subject in various parts of Europe and America, as well as in the East. The cases adduced to illustrate the text are, however, presented as types of hundreds of others obtainable from equally reputable sources, and to be found in the works of various trustworthy authorities, the titles of which can be seen in the Bibliography at the end of this volume.

The *London Review* for July, 1791, p. 40, referring to " An Essay on Vital Suspension : Being an Attempt to Investigate and Ascertain those Diseases in which the Principles of Life are Apparently Extinguished," by a Medical Practitioner—observes, that this is one of many publications " written by physicians and surgeons, versed in medical science, and well skilled in anatomy, to demonstrate, beyond a possibility of contradiction, that there are many cases in which the human body has the appearance of death, and preserves it for a considerable time, without the reality; the vital principle being still unsubdued, and a restoration of all its powers and functions practicable by the administration, in due time, of proper means." The author of the pamphlet under review says, " It is a proof of the temerity and imbecility of human judgment, that we have too many instances on record, wherein even the most skilful physicians have erred in the decisions they have pronounced respecting the extinction of life."

Unfortunately, we appear to be no nearer the prevention of these terrible mistakes now than we were when the reviewer called attention to them a century ago. The imbecility of human judgment complained of exists now in an unmitigated degree. The appearance of death is generally taken for its reality: and the great mass of the inhabitants of this planet are hurried to their graves without (except in a comparatively few cases of drowning or poisoning) the application of any serious efforts at restoration, and without waiting for unequivocal signs of dissolution.

Whether the risks of being buried alive are as great as those declared by some of the authorities quoted in this volume, must be left to the reader to determine for himself; but that they are considerable there appears little room for doubt by those who have taken the trouble to inquire into the facts. How often is the reader shocked by reading narratives in his daily or weekly newspaper of persons either buried alive, or of those in a state of suspended animation, but diagnosed and duly certificated by the attending doctor as dead, who have returned to consciousness during the funeral rites or at the grave itself.

The *Lancet* has borne frequent testimony to these disasters, some of which are quoted in this volume ; and, just as we are writing the closing chapter, the leading medical journal, in its issue of September 12, 1896, p. 785, records the following from its Cork correspondent as having occurred at Little Island, Ireland, which, the writer says, is thoroughly vouched for :—

"A child of four years of age contracted (typhoid) fever, and to all ordinary appearances died. The time of the funeral was appointed, and friends were actually on their way to attend it. When the supposed corpse was about to be removed from the bed to the coffin, signs of animation were exhibited. The services of the medical man were again requisitioned, and the child, opportunely rescued from such a terrible death, is now progressing satisfactorily."

Amongst the headings of paragraphs taken from recent papers lying before me are the following :— "Buried Alive," "A Gruesome Narrative," "Restored to Life in a Mortuary," "Premature Burial," "The Dead Alive," "Buried Alive," "Sounds from Another Coffin," "Mistaken for Dead," "A Lady Nearly Buried Alive," "Revivification After Burial," "A Woman's Awful Experience," "Bolt Upright in His Coffin," "Almost Buried while Alive," "A Woman Buried Alive," "The Corpse Sat Up," "Alive in Her Coffin," "Seemed to Rise from Death," "Escaped Burial Alive," "Revival at a Wake," "Snatched from Death at the Graveside," "Laid Out, but not Dead," "Alive in His Grave," "Interment before Death," "Came to Life in the Coffin," "Corpse Seems to Live," "The Corpse Moved," etc.

According to the "London Manual and Municipal Year Book," 1896-97, there are over four hundred public authorities at work in governing London, who spend over twelve million pounds a year, and from other sources it is said that seven millions a year are collected in the Metropolis for charitable purposes, and yet there are no officials, associations, or insurance companies to safeguard the people either in this wealthy Metropolis or in any part of the United Kingdom against one of the most terrible physical calamities that can overtake any member of the human family.

The Registrar-General's Decennial Supplement for 1881-90, published this year (1896), includes a "Life Table" furnishing the expectations of life in England and Wales. It appears that the death-rate has fallen from 21·3 in the decade ending 1880 to 19·0 per thousand living in that ending 1890. The expectation of life at birth, according to the actuary's standard in the decade 1871-80, was 41·3 years for males, and 44·6 years for females. This has been increased, as shown in the "Life Table" 1881-90, to 43·6 for males, and 47·2 for females, mainly through sanitary amelioration. A perceptible increase, the author believes, could be shown if steps were taken to restore still-born children, who constitute about five per cent. of births, and if the same trouble were adopted to restore the apparently dead from other diseases (which are sometimes only crises of repose after wasting disease) as is generally taken with respect to those accidentally poisoned or drowned. Besides reducing the mortality and increasing the expectation of life, such a reform would greatly diminish the appalling suffering of those

who, through our apathy and ignorance, are, under our
present system of *laisses faire*, consigned to precipitate
interment, and would bring tranquillity of mind to those
who are haunted all their lives through fear of such
a catastrophe. Why we should limit our efforts at
restoration of those apparently dead to a few cases has
never been shown, and is surely a serious oversight,
which should be remedied without delay.

Dr. Hartmann, in " Premature Burial," observes—" As
by cleaning a dusty watch the watchmaker causes the
hindrances to be removed which prevented the energy
stored up in the watch from setting the clockwork in
motion, so, in cases of apparent death from catalepsy,
asphyxia, syncope, and other diseases causing obstacles
to the manifestation of the life-energy in the body, these
obstacles may be removed by appropriate means, such
as are known to many intelligent physicians, and the
energy of life being latent in the physical form may be
enabled to manifest itself again when the harmony of
the organism has been sufficiently restored, even after
the heart has entirely ceased to beat."

Dr. A. Fothergill says :—" Since no one, from prince to
peasant, can at all times be secure from these dreadful
disasters, which suddenly suspend vital action ; and
since medical practitioners themselves are not exempt,
it surely becomes them to use every exertion to *improve* the art of *restoring animation.* May each
progressive step in this interesting path of science
tend to that great object! and may every laudable
attempt undertaken with that benevolent view enable
us with more certainty to preserve life and to diminish
the sum of human infelicity !"

It is regrettable that medical practitioners, neither in this nor in any of the Continental states, except, possibly, a few in Germany, have been trained to distinguish apparent from real death ; and when a case of death-trance occurs, they certify to actual death, and the unfortunate person is interred in a strong coffin, which effectually conceals the tragedy following resuscitation. Moreover, the ordinary practitioner, both in England and the United States, considers himself exonerated from blame when he thus follows the traditions and practice of the heads of his profession. Personally, he has neither the time, opportunity, or inclination to study the abnormal phenomena of trance, catalepsy, or hypnotisation, and thus the evil of live sepulture is perpetuated from generation to generation.

SUMMARY OF CONCLUSIONS.

(1) An examination of both the historical and modern cases of trance, catalepsy, and other death-counterfeits shows that nothing is more uncertain than the so-called signs of death, and that in all countries and in all ages many persons supposed by their attendant physicians and relations to be dead have revived, while the cases are as numerous and the danger as great at the present day as at any previous period.

(2) That the risk of premature burial is especially serious in France, in Spain and Portugal, in the west of

Ireland, in both European and Asiatic Turkey, and in India ; also amongst the Jews, where both the Jewish law and ancient customs enjoin burial within a few hours of death, and for similar reasons in all Oriental countries; and in the Southern States of North America.

(3) That the various signs which are supposed to indicate death, such as the cessation of respiration and of cardiac action, a pale, waxy and death-like appearance, the stiffening of the limbs, or *rigor mortis*, insensibility to cutaneous excitation, the departure of heat from the body, are singly and collectively illusory; the only safe and infallible test of dissolution being the manifestation of putrefaction in the abdomen.

(4) That medical death-certificates have been shown by various witnesses before the Select Parliamentary Committee of Inquiry of 1893-94 to be often misleading as to the cause of death, and inconclusive as to the fact of death. Any compulsory extension of the death-certification system in the present imperfect state of medical knowledge would only partially meet the necessities of the case, and might have the effect of crystallising a defective system into perfunctory routine. A certain safeguard would, however, be provided if the law made it binding on medical practitioners to set forth on the death-certificate a precise statement of indications showing that dissolution has actually occurred.

(5) That the only safe and effective method of reform is the establishment of appropriately designed waiting mortuaries, such as are provided at Munich, Weimar, Stuttgart, and other German cities, with qualified atten-

dants and appliances for resuscitation, and where doubtful cases of death (and all are doubtful in which decomposition has not clearly manifested itself) can be deposited until the fact of death is unequivocally established.

(6) That premature burial in civilised countries is mainly possible owing to the fact that instruction in the phenomena of trance, catalepsy, syncope, and other forms of suspended animation is not systematic in the medical schools in any country, and the means of prevention are therefore practically unknown. This omission should be immediately remedied by the inclusion of the subject at the appropriate place in the medical curriculum, and in the examination for degrees.

(7) That, inasmuch as a radical change in the methods of treating the dead or supposed dead is extremely urgent, and legislation with an overworked Parliament in England and apathetic State Legislatures in America will probably be delayed, the authors recommend, as a preliminary measure of protection, the formation of associations for the prevention of premature burial amongst their members, as in some cities in France, Austria, and the United States, or the alternative plan of engrafting such an obligation of prevention upon existing associations, clubs, and insurance companies established for other purposes.

If the foregoing conclusions are established, the need for immediate action is urgent and imperative, and the prompt intervention of Parliament should be at once invoked. May we hope for the cordial co-operation of all classes and of all sections on a question in which the whole community have a deep and vital interest, and on which procrastination will certainly be fatal to some of its members. It is not an academic question, but one of the gravest practical character, the earnest consideration and treatment of which cannot be neglected with impunity.

APPENDICES.

APPENDIX A.

HISTORICAL CASES OF RESTORATION FROM APPARENT DEATH.

FROM the time of Kornmann, Terilli, and Zacchia (see "Bibliography," seventeenth century), certain notable instances, from old authors, of restoration from apparent death have been cited, with a good deal of uniformity, in essays or theses on this subject. One of the most convenient (to English readers) of these compilations is to be found in an anonymous essay, "The Uncertainty of the Signs of Death," Dublin, 1748 (printed by George Faulkner), from which the following extracts are taken *verbatim* :—

Plutarch informs us that a certain person fell from an eminence, but did not show the least appearance of any wound, for, three days after, he suddenly resumed his strength, and returned to life as his friends were conveying him to the grave.

Asclepiades, a celebrated physician, on his return from his country seat, met a large company conveying a corpse to the grave. A principle of curiosity induced him to ask the name of the deceased person, but grief and sorrow reigned so universally that no one returned him answer; upon which, approaching the corpse, he found the whole of it rubbed over with perfumes, and the mouth moistened with precious balm, according to the custom of the Greeks; then carefully feeling every part, and discovering latent signs of life, he forthwith affirmed that the person was not dead, and the person was saved.—*Celsus ii., 6, " De re Medica."*

In the tenth book of Plato's Republic is related the story of one Er, an Armenian, who was slain in battle. Ten days after, when the surviving soldiers came with a view to inter the dead, they found all the bodies corrupted except his; for which reason they conveyed him to his own house in order to inter him in the usual manner. But two days after, to the great surprise of all present, he returned to life when laid on the funeral pile. Quenstedt remarks upon this case, which he took from

22

Kornmann's treatise "De Miraculis Mortuorum," "That the soul sometimes remains in the body when the senses are so fettered, and, as it were, locked up, that it is hard to determine whether a person is dead or alive." Pliny in his "Natural History," book vii., chap. 52, which treats of *those who have returned to life when they were about to be laid in the grave*, tells us that Acilius Aviola, a man of so considerable distinction that he had formerly been honoured with the consulship, returned to life when he was upon the funeral pile ; but, as he could not be rescued from the violence of the flames, he was burnt alive. The like misfortune also happened to Lucius Lamia, who had been praetor. These two shocking accidents are also related by Valerius Maximus. Celius Tubero had a happier fate than his two fellow-citizens, since, according to Pliny, he discovered the signs of life before it was too late. His state, however, was far from eligible, since, being laid on the funeral pile, he stood a fair chance of being exposed to the like misfortune. Pliny, from the testimony of Varro, adds that when a distribution of land was making at Capua, a certain man, when carried a considerable way from his own house in order to be interred, returned home on foot. The like surprising accident also happened at Aquinum. The last instance of this nature related by the author occurred at Rome, and Pliny must, no doubt, have been intimately acquainted with all its most minute circumstances, since the person was one Cerfidius, the husband of his mother's sister, who returned to life after an agreement had been made for his funeral with the undertaker, who was probably much disappointed when he found him alive and in good health.

These examples drawn from Roman history greatly contribute to establish the uncertainty of the signs of death, and ought to render us very cautious with respect to interments.

Greece and Italy are not the only theatres in which such tragical events have been acted, since other countries of Europe also furnish us with instances of a like nature. Thus, Maximilian Misson, in his "Voyage Through Italy," tome i, letter 5, tells us—

" That the number of persons who have been interred as dead, when they were really alive, is very great in comparison with those who have been happily rescued from their graves ; for, in the town of Cologne, Archbishop Geron—according to Albertus Krantzïus—was interred alive, and died for want of a seasonable releasement."

It is also certain that in the same town the like misfortune happened to Johannes Duns Scotus, who in his grave tore his hands and wounded his head. Misson also relates the following :—

" Some years ago the wife of one, Mr. Mervache, a goldsmith of Poictiers, being buried with some rings on her fingers, as she had desired when dying, a poor man of the neighbourhood, being apprised of that circumstance, next night opened the grave in order to make himself master of the rings, but as he could not pull them off without some violence, he in the attempt waked the woman, who spoke distinctly, and complained of the injury done her. Upon this, the robber made his escape. The woman, now roused from an apoplectic fit, rose from her coffin, returned to her own house, and in a few days recovered a perfect state of health."

What induced Misson to relate these histories was a certain piece of painting preserved in the Church of the Holy Apostles at Cologne, in order to keep up the memory of a certain accident, which that traveller relates in the following manner :—

"In the year 1571, the wife of one of the magistrates of Cologne being interred with a valuable ring on one of her fingers, the grave-digger next night opened the grave in order to take it off, but we may readily suppose that he was in no small consternation when the supposed dead body squeezed his hand, and laid fast hold of him, in order to get out of her coffin. The thief, however, disengaging himself, made his escape with all expedition ; and the lady, disentangling herself in the best manner she could, went home and knocked at her own door, where, after shivering in her shroud, after some delay she was admitted by the terror-stricken servants ; and, being warmed and treated in a proper manner, completely recovered."

Simon Goubart, in his admirable and memorable histories, printed at Geneva in 1628, relates the following accident :—"A lady, whose name was Reichmuth Adoloh, was supposed to fall a victim to a pestilence, which raged with such impetuous fury as to cut off most of the inhabitants of Cologne. Soon after, however, she not only recovered her health, but also brought into the world three sons, who, in process of time, were advanced to livings in the Church."

"The town of Dijon, in Burgundy, was, in the year 1558, afflicted with a violent plague, which cut off the inhabitants so fast that there was not time for each dead person to have a separate grave ; for which reason large pits were made and filled with as many bodies as they could contain. In this deplorable conjuncture, Mrs. Nicole Tentillet shared the common fate, and after labouring under the disorder for some days, fell into a syncope so profound that she was taken for dead, and accordingly buried in a pit with the other dead bodies. The next morning after her interment she returned to life, and made the strongest efforts to get out, but was held down by the weight of the bodies with which she was covered. She remained in this wretched condition for four days, when the grave-diggers took her out and carried her to her own house, where she recovered perfectly." Following this case, that of a labouring man of Courçelles, near Neuchâtel, is narrated. He fell into so profound syncope that he was taken for dead ; but the persons who were putting him into his grave without a coffin. perceived some motion in his shoulders, for which reason they carried him to his own home, where he perfectly recovered. This accident laid the foundation for his being called the ghost of Courcelles.

"A lawyer of Vesoul, a town of Franche-Comté, near Besançon, so carefully concealed a lethargy, to which he was subject, that nobody knew anything of his disorder, though the paroxysms returned very frequently. The motive which principally induced him to this secrecy was the dread of losing a lady to whom he was just about to be married. Being afraid, however, lest some paroxysm should prove fatal to him, he communicated his case to the Sheriff of the town, who, by virtue of his office, was obliged to take care of him if such a misfortune should happen. The marriage was concluded, and the lawyer for a considerable time enjoyed a perfect state of health, but at last was seized with so violent a paroxysm of the disease that his lady, to whom he had not revealed the secret, not doubting his death, ordered him to be put in his coffin. The Sheriff, though absent when the paroxysm seized him, luckily returned in time to preserve him ; for he ordered the interment to be delayed. and the lawyer, returning to life, survived the accident sixteen years."

Another case is that of a certain person who was conveyed to the church in order to be interred, but one of his friends sprinkling a large quantity of holy water on his face, which was covered, he not only returned to life, but also resumed a perfect state of health.

This writer subjoins other histories of persons who, being interred alive, have expired in their graves and tombs, as has afterwards been discovered by various marks made, not only in their sepulchres, but also in their own bodies. He in a particular manner mentions a young lady of Auxbourg, who, falling into a syncope, in consequence of a suffocation of the matrix, was buried in a deep vault, without being covered with earth, because her friends thought it sufficient to have the vault carefully shut up. Some years after, however, one of the family happened to die; the vault was opened, and the body of the young lady found on the stairs at its entry, without any fingers on the right hand.

It is recorded in "Tr. de Aere et Alim. defect.," cap. vii., that a certain woman was hanged, and in all appearances was dead, who was nevertheless restored to life by a physician accidentally coming in and ordering a plentiful administration of sal ammoniac.

Another case of hanging is the story of Anne Green, executed at Oxford, December 14, 1650. She was hanged by the neck for half an hour, some of her friends thumping her on the breast, others hanging with all their weight upon her legs, and then pulling her down again with a sudden jerk, thereby the sooner to despatch her out of her pain. After she was in her coffin, being observed to breathe, a lusty fellow stamped with all his force on her breast and stomach to put her out of pain. But by the assistance of Dr. Petty, Dr. Willis, Dr. Bathurst, and Dr. Clark, she was again brought to life.

Kornmann, in his treatise " De Miraculis Mortuorum," relates the following history :—"Saint Augustine, from Saint Cirille, informs us that a Cardinal of the name of Andrew having died in Rome in the presence of several bystanders, was next day conveyed to the church, where the Pope and a body of the clergy attended service in order to do honour to his memory. But to their great surprise, after some groans, he recovered his life and senses. This event was at the time looked upon as a miracle, and ascribed to Saint Jerome to whom the Cardinal was greatly attached."

The following account seems more to resemble a miracle, though we do not find that it was looked upon as such :—" Gocellinus, a young man, and nephew to one of the Archbishops of Cologne, falling into the Rhine, was not found for fifteen days after, but was discovered to be alive as he lay before the shrine of Saint Guibert."

Persons curious or incredulous upon the dangers of precipitate burials may, for their satisfaction, have recourse to the medical observations of Forestus ; those of Amatus Lusitanus ; the chirurgical observations of William Fabri ; the treatise of Levinus Lemnius on the secret miracles of Nature ; the observations of Schenkins ; the medico-legal questions of Paul Zacchias ; Albertinus Bottonus's treatise of the Disorders of Women ; Terilli's treatise on the Causes of Sudden Death ; Lancisi's treatise Concerning Deaths, and Kornmann's treatise on the Miracles of the Dead. These authors furnish us with a great variety of the most palpable and flagrant instances of the uncertainty of the signs of death. As examples of the possibility of even great anatomists being imposed upon by these fallacious signs, the two following accidents are given :—

" Andreas Vesalius, successively first physician to Charles the Fifth and his son Philip the Second of Spain, being persuaded that a certain Spanish gentleman, whom he had under management, was dead, asked liberty of his friends to lay open his body. His request being granted, he no sooner plunged his dissecting-knife in the body than he observed signs of life in it, since, upon opening the breast, he saw the heart palpitating. The friends of the deceased, horrified by the accident, pursued Vesalius as a murderer ; and the judges inclined that he should suffer as such. By the entreaties of the King of Spain, he was rescued from the threatening danger, on condition that he would expiate his crime by undertaking a voyage to the Holy Land."

The account of the accident that befell the other anatomist is taken from Terilli, and runs as follows :—

"A lady of distinction in Spain, being seized with an hysteric suffocation so violent that she was thought irretrievably dead, her friends employed a celebrated anatomist to lay open her body to discover the cause of her death. Upon the second stroke of the knife she was roused from her disorder, and discovered evident signs of life by her lamentable shrieks extorted by the fatal instrument. This melancholy spectacle struck the bystanders with so much consternation and horror that the anatomist, now no less condemned and abhorred than before applauded and extolled, was forthwith obliged to quit not only the town but also the province in which the guiltless tragedy was acted. But though he quitted the now disagreeable scene of the accident, a groundless remorse preyed upon his soul, till at last a fatal melancholy put an end to his life."

Physicians of the earlier ages knew that there were disorders which so locked up or destroyed the external senses that the patients labouring under them appeared to be dead. According to Mr. Le Clerc, in his "History of Medicine," Diogenes Laertius informs us "that Empedocles was particularly admired for curing a woman supposed to be dead, though that philosopher frankly acknowledged that her disorder was only a suffocation of the matrix, and affirmed that the patient might live in that state (the absence of respiration) for thirty days."

Mr. Le Clerc, in the work already quoted, tells us that "Heraclides of Pontus wrote a book concerning the causes of diseases, in which he affirmed that a patient is without respiration in certain disorders for thirty days, and that they appeared dead in every respect, except corruption of the body."

To these authorities we may add that of Pliny, who, after mentioning the lamentable fate of Aviola and Lamia, affirms—"That such is the condition of humanity, and so uncertain the judgment men are capable of forming of things, that even death itself is not to be trusted to.'

Colerus, in "Oeconom." part vi., lib. xviii., cap. 113, observes, "That a person as yet not really dead may, for a long time, remain apparently in that state without discovering the least signs of life; and this has happened in the times of the Plague, when a great many persons interred have returned to life in their graves." Authors also inform us that the like accident frequently befalls women seized with a suffocation of the matrix (hysteria).

Forestus, in "Obs. Med.," l. xvii., obs. 9, informs us—"That drowned persons have returned to life after remaining forty-eight hours in the water; and sometimes women, buried during a paroxysm of the hysteric passion, have returned to life in their graves; for which reason it is forbidden in some countries to bury the dead sooner than seventy-two hours after death." This precaution of delaying the interment of persons thought to be dead is of a very ancient date, since Dilberus, in "Disput. Philol.," tome i., observes that Plato ordered the bodies of the dead to be kept till the third day. *in order to be satisfied of the reality of death.*

The burial customs of the ancients often included steps that were taken as a precaution against mistaking the living for the dead. Indeed the fear of such an accident seems to have always been entertained as a thing liable to occur in every case of seeming death. The embalming process employed by the Egyptians was a surgical test of the kind. The abdomen was first opened in order to remove the intestines, and some startling experiences must have been had in consequence of the incisions required for this operation, because it was customary for the friends and relatives of the deceased to throw stones at the persons employed in embalming as soon as the work was over, owing to the horror with which they were struck upon witnessing what must have been at times a cruel proceeding.

The funeral ceremonies used in the Caribbee Islands are, in a great measure, conformable to reason. They wash the body, wrap it up in a cloth, and then begin a series of lamentations and discourses calculated to recall the deceased to life, by naming all the pleasures and privileges he has enjoyed in the world, saying over and over again, "How comes it, then, that you have died?" When the lamentations are over, they place the body on a small seat, in a grave about four or five feet deep, and for ten days present aliments to it, entreating it to eat. Then, convinced that it would neither eat nor return to life, they, for its obstinacy, throw the victuals on its head, and cover up the grave. It is evident from the practices of this people that they wait so long before they cover the body with earth, because they have had instances of persons recalled to life by these measures.

Lamentations of a similar kind were employed by the Jews and Romans, as well as by the ancient Prussians and the inhabitants of Servia, founded doubtless upon similar experiences.

The Thracians, according to Herodotus, kept their dead for only three days, at the end of which time they offered up sacrifices of all kinds, and, after bidding their last adieu to the deceased, either burned or interred their bodies.

According to Quenstedt, the ancient Russians laid the body of the dead person naked on a table, and washed it for an hour with warm water. Then they put it into a bier, which was set in the most public room in the house. On the third day they conveyed it to the place of interment. where the bier, being opened, the women embraced the body with great lamentations. Then the singers spent an hour in shouting and making a noise in order to recall it to life; after which it was let down into the grave and covered with earth. So that this people used the test of warm water, that of cries, and a reasonable delay, before they proceeded to the interment.

In the laws and history of the Jews, there is but one regulation with respect to interment (in the twenty-first chapter of Deuteronomy), where the Jewish legislator orders persons hanged to be buried the same day. From this, one is led to infer that the funeral ceremonies, as handed down from Adam, were otherwise perfect and unexceptionable. The bier used by the Jews, on which the body was laid, was not shut at the top, as our coffins are, as is obvious from the resurrection of the Widow of Nain's son, recorded in the seventh chapter of Luke, where these words occur :—
"And he came and touched the bier, and they that bare him stood still. And he said, Young man, I say unto thee, Arise ; and he that was dead sat up and began to speak."

Gierus and Calmet inform us that the body, before its interment, lay for some days in the porch or dining-room of the house. According to Maretus, it was probably during this time that great lamentations were made, in which the name of the deceased was intermixed with mournful cries and groans.

Mr. Boyer, member of the Faculty at Paris, observes that such lamentations are still used by the Eastern Jews, and even by the Greeks who embrace the articles of the Greek Church. These people hire women to weep and dance by turns round the body of the dead person, whom they interrogate with respect to the reasons they had for dying.

Lanzoni, a physician of Ferrara, informs us that "when any person among the Romans died, his nearest relatives closed his mouth and eyes, and when they saw him ready to expire, they caught his last words and sighs. Then calling him aloud three times by his name, they bade him an eternal adieu." This ceremony of calling the name of the dying person was called Conclamation, a custom that dates prior to the foundation of Rome, and was only abolished with paganism.

Propertius acquaints us with the effect they expected from the first Conclamation—since there were several of them. He introduces Cynthia as saying, "Nobody called me by my name at the time my eyes were closing, and I should have enjoyed an additional day if you had recalled me to life."

Conclamations were made also by trumpets and horns, blown upon the head, into the ears, and upon neck and chest, so as to penetrate all the cavities of the body, into which, as the ancients imagined, the soul might possibly make her retreat.

Quenstedt and Casper Barthius, in "Advers.," lib. xxxvii., ch. 17, tell us that it was customary among the ancients to wash the bodies of their dead in warm water before they burned them, "that the heat of the water might rouse the languid principle of life which might possibly be left in the body."

By warm water we are to understand boiling water, as is obvious from the copious steam arising from the vessel represented in pieces of statuary in such instances: as also from the Sixth Book of Virgil's "Æneid"— "Some of the companions of Æneas, with boiling water taken from brazen vessels, wash the dead body, and then anoint it."

The Romans, as Lanzoni informs us, kept the bodies of the dead seven days before they interred them ; and Servius, in his commentary on Virgil, tells us "that on the eighth day they burned the body, and on the ninth put its ashes in the grave." Polydorus and Alexander ab Alexandro are also of opinion that the Romans kept the dead seven days ; and Gierus affirms that they sometimes did not bury them till the ninth : but it is easy to believe that they deviated from the most universal custom when evident and incontestable marks of death rendered it safe to inter before the usual time. Alexander ab Alexandro also observes that it was customary among the Greeks to keep the bodies of their dead seven days before they put them on the funeral pile.

It would have, perhaps, been sufficient to have kept the bodies of the dead seven days, or nine, or till putrefaction evinced the certainty of death ; but the Romans carried their circumspection farther, since, to use the words of Quenstedt, "Those who were employed in watching the dead now and then began their conclamations, and all at once called the dead person aloud by his name, because, as Celsus informs us, the principle of life is often thought to have left the body when it still remains in it : for which reason conclamations were made, in order, if possible, to rouse it and excite it."

If our senses are so imperfect that the signs of life may escape them ; if the languid state of the sensitive powers, or the origin of the nerves, is such that the most painful chirurgical operations are sometimes insufficient to put the spirits in motion ; if the duration of a perfect insensibility for a considerable number of days is a precarious and uncertain mark of death ; and if situations, apparently the most inconsistent with life, for a considerable time amount only to strong presumptions that life is destroyed, we ought, with Mr. Winslow and a great many other celebrated authors, to conclude that a beginning of putrefaction is the only certain sign of death.

Mr. Winslow evidently proves that the most cruel chirurgical operations are sometimes insufficient to ascertain death. From these observations we can but conclude—(1) That it is to no purpose to use the most cruel chirurgical operations ; and (2) that it is necessary to abstain from such as may prove mortal to the patient. Mr. Winslow is indeed so far from recommending operations of the last mentioned kind, that he calls it rash to plunge a long needle under the nail of an apoplectic patient's toe.

But if Mr. Winslow thinks it rash to make a simple puncture in a nervous part, we ought, surely, not to entertain a favourable notion of the large and enormous incisions made in dissections. Those, indeed, who are dissected run no risk of being interred alive. The operation is an infallible means to secure them from so terrible a fate. This is one advantage which persons dissected have over those who are, without any further ceremony, shut up in their coffins.

———

In the appendix to the second edition of Dr. Curry's "Observations on Apparent Death" several instances of a similar kind are added, and amongst others the case of William Earl of Pembroke, who died April 30, 1630. When the body was opened in order to be embalmed, he was observed, immediately after the incision was made, to lift up his hand. This is capped by the incident of Vesalius already given.

"A correspondent of the late Dr. Hawes assures us that there was then living in Hertfordshire a lady of an ancient and honourable family whose mother was brought to life after interment by the attempt of a thief to steal a valuable ring from her finger. (See Reports of the Royal Humane Society for 1787-88-89, p. 77.) Whether it was the same or not I cannot say, but Lady Dryden, who resided in the southern part of Northamptonshire, in consequence of some such event having occurred in her family expressly directed in her will that her body should have the throat cut across previous to interment ; and to secure this bequeathed fifty pounds to an eminent physician, who actually performed it."—*Ibid., p. 106.*

Dr. Elliotson refers to a case of a female who was pronounced to be dead. Her pulse could not be felt, and she was put into a coffin ; and, as the coffin lid was being closed, they observed a sweat break out, and thus saw that she was alive. She recovered completely, and then stated that she had been unable to give any signs of life whatever ; that she was conscious of all that was going on around her ; that she heard everything ; and that when she found the coffin lid about to be put on, the agony was dreadful beyond all description, so that it produced the sweat seen by the attendants.

DEATH-TRANCE.

In two cases related by the late Mr. Braid, of Manchester, "the patients remained in the horrible condition of hearing various remarks about their death and interment. All this they heard distinctly without having the power of giving any indication that they were alive, until some accidental abrupt impression aroused them from their lethargy, and rescued them from their perilous situation. On one of these occasions, what most intensely affected the feelings of the entranced subject, as she afterwards communicated to my informant, was hearing a little sister, who came into the room, where she was laid out for dead, exulting in the prospect, in consequence of her death, of getting possession of a necklace of the deceased." In another instance, the patient remained in a cataleptic condition for fourteen days. During this period, the visible signs of vitality were a slight degree of animal heat and appearance of moisture when a mirror was held close to her face. But although she had no voluntary power to give indication by word or gesture, nevertheless she heard and understood all that was said and proposed to be done, and suffered the most exquisite torture from various tests applied to her. . . . There is hardly a more interesting chapter in the records of medical literature than the history of well-authenticated cases of profound lethargy or death-trance. Most of the reported cases in which persons in a state of trance are stated to have been consigned to the horrors of a living burial may possibly be apocryphal. Still, on the other hand, there are unquestionably too many well-substantiated instances of the actual occurrence of this calamity, the horrors of which no effort of the imagination can exaggerate, and for the prevention of which no pains can be excessive and no precaution superfluous.

The following is taken from "Memorials of the Family of Scott, of Scott's Hall, in the County of Kent, with an Appendix of Illustrative Documents," by James Benat Scott, F. S. A., London, 1876, page 225 :—

"Robert Scott, Esq., tenth (but sixth surviving) son of Sir Thomas Scott, of Scot's-Hall, Knight, married Priscilla, one of the daughters of Sir Thomas Honywood, of Elmsmere, Knight, by whom he had nine children. Remarkable accidents happened to the said Robert Scott and Priscilla, his wife, before their marriage, at their marriage, and after their marriage, before they had children. At their marriage, which was in or about the year 1610, the said Robert Scott having forgot his wedding ring when they were to be married, the said Priscilla was married with a ring with death's head upon it.

" Within a short time after they were married, the said Robert Scott, and Priscilla, his wife, sojourning with Sir Edward at Austenhanger, the said Robert Scott, about Bartholomewtide, fell sick of a desperate malignant fever, and was given over for dead by all, insomuch as that he was laid forth, the pillows pulled from under him, the curtains drawn, and the chamber windows set open, and ministers spoke to to preach the funeral service, and a book called for his funeral that was to have been kept at Scot's Hall, where Sir John Scott the eldest brother then lived. At night he was watched with by his own servant, named Robins, and another servant in the house, and about midnight they, sitting together by the fire in the chamber, the said Robins said to the other, ' Methinks my master should not be dead, I will go and try,' and presently starting up went to the bedside where his master laid, and hallooed in his ear, and laid a feather to his nostrils, and perceived that he breathed, upon which he called them up in the house, and they warmed clothes and rubbed him, and brought him to life again.   He lived afterwards to be upwards of seventy-two years of age, and to have nine children.

" Another remarkable passage was that his wife, Priscilla, being then very sick also, they told her that he was dead. She answered that she did not believe that God would part them so soon. The said Priscilla, when born, was laid for dead, no one minding her, but all the women went to help her mother, who was then like to die after her delivery ; but at last an old woman, taking the child in her arms, carried it downstairs, and using means, brought her to life. The other women, missing the child, and hearing the old woman had carried her down to get life in her, laughed at her, as thinking it impossible to bring the child to life ; but in a little time she brought it into the chamber, to the amazement of them all, and said she might live to be an old woman ; and so she did to the age of fifty-two, and had nine children."

The following cases are from Mrs. Crowe's " Night Side of Nature," pp. 133-136 :—

" Dr. Burns mentions a girl at Canton, who lay in a trance, hearing every word that was said around her, but utterly unable to move a finger. She tried to cry out, but could not, and supposed that she was really dead. The horror of finding that she was about to be buried at length caused a perspiration to appear on her skin, and she finally revived. She described that she felt that her soul had no power to act upon her body, and that it seemed to be *in her body and out of it at the same time.*"

" Lady Fanshawe related the case of her mother, who being sick of a fever, her friends and servants thought her deceased, and she lay in that state for two days and a night ; but Mr. Winslow, coming to comfort my father, went into my mother's room, and, looking earnestly into her face, said, 'She was so handsome, and looked so lovely, that he could not think her dead,' and, suddenly taking a lancet out of his pocket, he cut the sole of her foot, which bled: upon this he immediately caused her to be removed to the bed again, and she opened her eyes, after rubbing and other restorative means, and came to life."

"On the 10th of January, 1717, Mr. John Gardner, a minister at Elgin, fell into a trance, and being to all appearances dead, he was put into a coffin and on the second day was carried to the grave. But fortunately a noise being heard, the coffin was opened, and he was found alive and taken home again, where, according to the record, 'he related many strange and amazing things which he had seen in the other world.'"

Under the head of "Suspended Animation: Cases of Recovery, etc.," the Report of the Royal Humane Society for 1816-17, pp. 48-50, copies the following:—"A young lady, an attendant on the Princess of ——, after having been confined to her bed for a great length of time with a violent disorder, was at last to all appearances deprived of life. Her lips were quite pale, her face resembled the countenance of a dead person, and her body became cold.

"She was removed from the room in which she died, was laid in a coffin, and the day of her funeral was fixed on. The day arrived, and, according to the custom of the country, funeral songs and hymns were sung before the door. Just as they were about to nail on the lid of the coffin, a slight perspiration was observed to appear on the surface of her body. It grew greater every moment, and at last a kind of convulsive motion was observed in the hands and feet of the corpse. A few moments after, during which time fresh signs of returning life appeared, she at once opened her eyes, and uttered a pitiable shriek. Physicians were quickly procured, and in the course of a few days she was considerably restored, and is probably alive at this day."

The description which she herself gave of her situation is extremely remarkable, and forms a curious and authentic addition to psychology:—

"She said it seemed to her, as if in a dream, that she was really dead; yet she was perfectly conscious of all that happened around her in this dreadful state. She distinctly heard her friends speaking, and lamenting her death at the side of her coffin. She felt them pull on the dead-clothes and lay her in it. This feeling produced a mental anxiety which was indescribable. She tried to cry, but her soul was without power and could not act on her body. She had the contradictory feeling as if she were in her body, and yet not in it, at one and the same time. It was equally impossible for her to stretch out her arms, or to open her eyes, or to cry, although she continued to do so.   The internal anguish of her mind was, however, at its utmost height when the funeral hymns began to be sung, and when the lid of the coffin was about to be nailed on.   The thought that she was to be buried alive was the first one which gave activity to her soul, and caused it to operate on her corporeal frame."

Related by Dr. Herz in the "Psychological Magazine," and transcribed by Sir Alexander Crichton in the introduction to his essay on "Mental Derangement." [2 vols., Lond., 1798.]

"One of the most frightful cases extant is that of Dr. Walker, of Dublin, who had so strong a presentiment on this subject, that he had actually written a treatise against the Irish custom of hasty burial. He, himself, subsequently died, as was believed, of a fever. His decease took place in the night, and on the following day he was interred. At this time, Mrs. Bellamy, the once-celebrated actress, was in Ireland; and, as she had promised him, in the course of conversation, that she would take care he should not be laid in the earth till unequivocal signs of dissolution had appeared, she no sooner heard of what had happened than she took measures to have the grave reopened; but it was, unfortunately, too late. Dr. Walker had evidently revived, and had turned upon his side; but life was quite extinct."

Mr. Horace Welby, in a chapter on "Premature Interment," says that "the Rev. Owen Manning, the historian of Surrey, during his residence at Cambridge University, caught small-pox, and was reduced by the disorder to a state of insensibility and apparent death. The body was laid out and preparations were made for the funeral, when Mr. Manning's father, going into the chamber to take a last look at his son, raised the imagined corpse from its recumbent position, saying, 'I will give my poor boy another chance,' upon which signs of vitality were apparent. He was therefore removed by his friend and fellow-student, Dr. Heberden, and ultimately restored to health."—*The Mysteries of Life and Death*, *pp. 115-116.*

A most conspicuous and interesting monument in St. Giles's Church, Cripplegate, London (where Cromwell was married and John Milton buried), is associated with a remarkable case of trance or catalepsy. In the chancel is a striking sculptured figure in memory of Constance Whitney, a lady of remarkable gifts, whose rare excellences are fully described in the tablet.   She is represented as rising from her coffin.

Welby, at p. 116, relates the story that she had been buried while in a state of suspended animation, but was restored to life through the cupidity of the sexton, which induced him to disinter the body to obtain possession of a valuable ring left upon her finger, which he concluded could be of no use to the wearer. A study of the facts of premature burial shows that the rifling of tombs and coffins to obtain valuables has in other instances revealed similar tragic occurrences.

The often-cited case of Mrs. Goodman, one of those recalled to life by the sexton's attempt to remove a ring from the finger, is thus related in the "History of Bandon," by George Bennett :—

Hannah, wife of Rev. Richard Goodman, vicar of Ballymodan, Bandon, from 1692 to 1737, fell into ill-health, and apparently died. Two or three days after her decease, the body was taken to Rosscarbery Cathedral, and there laid in the family fault of the Goodmans. The attempt of the sexton to recover a valuable diamond ring from the finger is said to have been made at an early hour the next morning. Much violence was used, so that the corpse moved, yawned, and sat up. The sexton having fled in terror, leaving his lantern behind and the church door open, the lady in her shroud made her way out of the vault and through the church to the residence of her brother-in-law, the Rev. Thomas Goodman, which was just outside the church-yard. Having been admitted after some delay and consternation, she was put to bed, and fell asleep soon after, her brother-in-law and his man-servant keeping watch over her until midday, when she awoke refreshed. She is said to have shown herself in the village in the afternoon, to have supped with the family in the evening, and to have set out for home on horseback next morning. She is said to have survived this episode for some years, and to have borne a son subsequent to it, who died at an advanced age at Innishannon, a village near Bandon.

In Smith's "History of Cork," vol. ii., p. 428, the same incident is thus mentioned :—" Mr. John Goodman, of Cork, died in January, 1747, aged about four score; but what is remarkable of him, his mother was interred while she lay in a trance, having been buried in a vault, etc. . . . . This Mr. Goodman was born some time after."

———

Mr. Peckard, Master of Magdalen College, Cambridge, in a work entitled "Further Observations on the Doctrine of an Intermediate State," mentions that Mrs. Godfrey, Mistress of the Jewel Office, and sister of the great Duke of Marlborough, is stated to have lain in a trance, apparently dead, for seven days, and was declared by her medical attendants to have been dead.   Colonel Godfrey, her husband, would not allow her to be interred, or the body to be treated in the manner of a corpse ; and on the eighth day she awoke, without any consciousness of her long insensibility.

The daughter of Henry Laurens, of South Carolina, the first President of the American Congress during the Revolutionary War, died when young of small-pox.   At all events a medical certificate pronounced her dead, and she was shrouded and coffined for interment.   It was customary in those days to confine the patient amidst red curtains with closed windows. After the certificate of death had been duly made out, the curtains were thrown back and the windows opened.   The fresh air revived the patient, who recovered and lived to a mature age.   This circumstance occasioned on her father so powerful a dread of living interment, that he directed by will that his body should be burnt, and enjoined on his children the performance of this wish as a sacred duty.

Bouchut in his "Signes de la Mort," p. 58, relates that the physician of Queen Isabella of Spain was treating a man during a dangerous illness, and as he went to see his patient one morning he was informed by the assistants that the man had died.   He entered, and found the body, in the habit of the Order of St. Francis, laid out upon a board.   Nothing daunted, he had him put back to bed in spite of the ridicule of those present, and the patient soon revived and fully recovered.

The following cases are from Köppen (see Bibliography, 1799) :—

Vienna, 1791.—A castle guard (*portier*) was in a trance for several days. His funeral was prepared, and he was placed in a coffin.   All at once he unexpectedly opened his eyes and called out, "Mother, where is the coffee ?"

Halle, 1753.—In the register of deaths, at St. Mary's Church, is the following entry :—"Shoemaker Casper Koch was buried, aged eighty-one years. Thirty years ago he had died, to all appearances, and was put in a coffin, when suddenly, when they were about to bury him, he recovered his consciousness."

Haag, Holland, 1785.—The son of a cook died, and while the coffin was being carried to the grave-yard, he was heard to knock.   On opening the coffin he was found alive.   He was taken home and was restored.

In the "Cyclopædia of Practical Medicine," edited by John Forbes, M.D., F.R.S., and others, 1847, vol. i., pp. 548-549, is the following :— " A remarkable instance of resuscitation after apparent death occurred in France, in the neighbourhood of Douai, in the year 1745, and is related by

Rigaudeaux, (*Journal des Sçavans*, 1749,) to whom the case was confided. He was summoned in the morning to attend a woman in labour, at a distance of about a league. On his arrival, he was informed that she had died in a convulsive fit two hours previously. The body was already prepared for interment, and on examination he could discover no indications of life. The os uteri was sufficiently dilated to enable him to turn the child and deliver by the feet. The child appeared to be dead also; but, by persevering in the means of resuscitation for three hours, they excited some signs of vitality, which encouraged them to proceed, and their endeavours were ultimately crowned with complete success. Rigaudeaux again carefully examined the mother, and was confirmed in the belief of her death; but he found that, although she had been in that state for seven hours, her limbs retained their flexibility. Stimulants were applied in vain; he took his leave, recommending that the interment should be deferred until the flexibility was lost. At five p.m. a messenger came to inform him that she had revived at half-past three. The mother and child were both alive three years after."

APPENDIX B.

RESUSCITATION OF STILL-BORN AND OTHER INFANTS.

THE danger of premature burial of still-born (apparently dead) infants is clearly shown by the following quotation from Tidy's " Legal Medicine," part ii., page 253, from tables given on the authority of the *British and Foreign Medical Review*, No. ii., p. 235, based on eight millions of births. "It would appear that from one in eighteen to one in twenty births are still-born. Dr. Lever found that the proportion in his three thousand cases was one in eighteen. So notorious is it that a large number of these deaths could be averted, that some legislation is urgently needed, requiring that still-borns, whose bodies weigh, say, not less than two pounds (the average weight about the sixth and seventh months at which children are viable), should not be buried without registration and a medical examination."

Many instances can be found in current medical literature of still-born infants that have been revived by artificial respiration. Such cases not infrequently revive without any means being employed for their resuscitation; but among the poor, who dispose of the new-born apparently dead in a hasty manner, they might be buried alive through carelessness. The use of mortuaries, where the seeming dead would be kept under observation until decomposition appears, would of course prevent such disasters.

Struve, in the Essay cited in the Bibliography (1802), says :—

" All still-born children should be considered as only apparently dead, and the resuscitative process ought never to be neglected. Sometimes two hours or more will elapse before reanimation can be effected. An ingenious man-midwife, says Bruhier, was employed for several hours in the revival of an apparently still-born child, and as his endeavours proved unavailing, he considered the subject really dead. Being, however, accidentally detained, he again turned his attention to the child, and by continuing the resuscitative method for some time it was unexpectedly restored to life " (p. 150).

The following is one of Struve's most striking cases :—

A Mr. E.—— called in 18—— to obtain a certificate of death for a still-born child of seven months' gestation. Arriving at the house, the doctor found the child laid upon a little straw and covered with a slight black shawl ; this was one p. m., and the child had been there since five a. m. It was icy cold, and there was no heart sound nor respiration, but there was a slight muscular twitching over the region of the heart. The child was immersed in a hot bath and artificial respiration employed, but for twenty minutes the case seemed hopeless ; then the eyes opened and after continued effort the respirations began, laborious and interrupted at first, then normal by degrees. The child was saved, and became an accomplished violinist.

The mortality and waste of infant life, particularly in large cities like Paris, London, Berlin, Vienna, and New York, is admitted by all investigators to be enormous. In France medical writers, in view of the small percentage of births to population, are waking up to the realisation that the State cannot afford the loss, and that, among other things, steps should be taken to resuscitate the still-born, so that none should be buried before unequivocal signs of death are manifested.[1] The premature abandoment of the still-born among the poorer classes in crowded cities is only too probable. There are also cases recorded which show a corresponding risk to infants who have survived their birth :—

The *British Medical Journal*, January 21, 1871, p. 71, gives the following case, under the heading, "Alive in a Coffin":—"Stories of this kind are generally very apocryphal; but the following reaches us from an authentic source. A child narrowly escaped being buried alive last week

---

[1] During the five years ending 1895 the population of France, where of all European countries premature burial is most in vogue, has increased by only 133, 819, or, leaving out the immigration of alien population, the increase is under 30,000. The population for all practical purposes may be regarded as stationary.

in Manchester. The infant's father had died, and was to be buried in Ardwick Cemetery. The day before the burial the infant was taken ill, and apparently died. A certificate of death was procured from a surgeon's assistant who had seen the child, and, to save expense, it was decided to place it in the same coffin with the father. This was done, and the next morning the bearers set off to the cemetery with their double burden; but before reaching the graveyard a cry was heard to issue from the coffin. The lid being removed, the infant was discovered alive and kicking. It was at once removed to a neighbour's house, but died eight hours afterwards.

The *British Medical Journal*, 1885, ii., p. 841, gives the following case, under the heading, "Death or Coma?"

"The close similarity which is occasionally seen to connect the appearance of death with that of exhaustion following disease, was lately illustrated in a somewhat striking manner. An infant seized with convulsions was supposed to have died about three weeks ago at Stamford Hill. After five days' interval, preparations were being made for its interment, when, at the grave's mouth, a cry was heard to come from the coffin. The lid was taken off, and the child was found to be alive ; it was taken home, and is recovering."

The following is from Tidy's "Legal Medicine," pt. i., p. 29 :—

"In a communication to the French Academy, Professor Fort mentions a child (*ætat*. three) having been resuscitated by artificial respiration continued for four hours, and not commenced until three and a half hours after its apparent decease.

"Ogston records one case of a child alive for seven hours, and a second case of a young woman alive for four hours, after they had been left as dead."

From the *Lancet*, April 22, 1882, p. 675 :—

"PREMATURE INTERMENT.

"A daily contemporary states that at the gates of the Avignon cemetery the parents of a child certified to have died of croup insisted on having the coffin opened to take a last look. The child was found breathing, and is expected to be saved."

The following letter to the editor of the *Lancet*, March 31, 1866, p. 360, illustrates the danger to which infants supposed to be dead are exposed, under one of our traditional customs :—

"LAYING-OUT OF DEAD INFANTS.

"Sir,—In your journal of last Saturday, among the 'Medical Annotations,' you notice the inquiry into the circumstances under which an infant, being still living and moving, was 'bandaged' beneath the chin, and 'laid-out' at St. Pancras Workhouse. Allow me to state that in the *Lancet*, vol. ii., 1850, a contribution from me 'On the Danger of Tying-up the Lower Jaw immediately after Supposed Death' was published. An infant, aged two months, was brought to me on a Friday with the lower jaw tied up by its mother, who asked for a certificate of death; but on my removing the bandage, the child began to show symptoms of vitality, and it lived until the following Monday.

C. J. B. ALDIS, M.D., F.R.C.P.

"Chester Terrace, Chester Square, March 26, 1866."

It is recorded that Dr. Doddridge showed so little signs of life at his birth that he was laid aside as dead, but one of the attendants, observing some signs of life, took the baby under her charge, and by her judicious treatment perfectly restored it.

Mr. Highmore, Secretary of the London Lying-in Hospital, confirmed (by a communication to the Royal Humane Society, April, 1816,) the statement of Mrs. Catherine Widgen, the matron of that excellent establishment, that, by a zealous perseverance in the means recommended by that Society, she had been the happy instrument of restoring from a state of apparent death in the space of *three years* no less than forty-five infants, who, but for her humane attention and indefatigable exertions, must have been consigned to the grave. Later on, Mrs. Widgen restored in one year twenty-seven apparently dead-born children—a striking instance of the truth of the remark of a celebrated writer (Osiander) that "the generality of infants, considered as still-born, are only apparently so; if, therefore, persons would persevere in their exertions to revive them, most of them might be restored."—*Report of the Royal Humane Society, 1816-17, pp. 52-54.*

"For these exertions the General Court adjudged the Honorary Medallion to Mrs. Widgen, and it was accordingly presented to her by His Royal Highness the Duke of Kent."—*Ibid., p. 52.*

[The question naturally suggests itself in this place: If the matron of such a noble institution as the above was able to save seventy-two apparently dead children from the grave in four years, how many of these poor little beings are consigned to the grave all over the world for lack of the "humane attention and indefatigable exertions," such as this skilful matron gave to those that came under her intelligent care?]

"RECURRENCE OF SUSPENDED ANIMATION.

"A child, who had a cough for some time, was suddenly attacked with difficulty of breathing, and *to all appearances died.* A medical gentleman immediately inflated the lungs, and by persisting in this for a considerable time, recovered the child. A similar state of suspended animation took place three or four times, and inflation was as often had recourse to with the same success ; but the attack, happening, unfortunately, to recur whilst the medical gentleman in whose family the case happened was from home, the proper measures were not taken, and the child was lost."—*Ibid., p. 140.*

" SHOCK FROM LIGHTNING.

" A boy was struck down by a flash of lightning near Hoxton (in the suburbs of London), and lay exposed to the rain at least an hour, until his companions carried him home on some boards, apparently dead—the body being stiff and universally cold, the fingers and toes contracted, and the countenance livid. He was stripped of his wet clothes, put in hot blankets, and bled twenty ounces. In half an hour, interrupted respiration commenced, without inflating the lungs ; in an hour more, regular pulsation and breathing were established, together with power of swallowing ; and in a week he was quite well."—*Ibid., p. 147.*

In the *Lancet*, 1884, vol. i., p. 922, W. Arnold Thompson, F.R.C.S.I., reports a case of resuscitation of a child delivered by the forceps, which was " apparently to myself [he says] and the nurse and relatives, a perfectly dead child, and with no signs of respiration or life about it. . . . My opinion was that the death was real and positive, but that, there being no actual disease present, and the blood still warm, the machinery of life was set going, and resuscitation followed as a consequence of suitable means being taken and persevered in without undue delay. In the future I do not intend to allow any still-born children to be put away without making strenuous efforts to restore vitality."

The *Lancet*, 1880, vol. ii., p. 582 :—In a discussion at the Royal Medical and Chirurgical Society upon Artificial Respiration in New-born Children, Dr. Roper related three cases in which the child was left for dead. " One of these occurred in the practice of Mr. Brown, of St. Mary Axe. The child was still-born in the absence of a medical man. It was taken to the surgery, and thence to the late Mr. Solly, who next day, in dissecting the body, found that the heart was still beating. A second instance was of a fœtus of five months and a half, which was set aside as dead, Dr. Roper attending the mother, who was suffering from hæmorrhage. He was astonished next day to find that this immature child, which had lain on the floor for eleven hours through a cold night, was breathing and its heart beating. . . ." Such examples show that the new-born have greater tenacity of life than is supposed.

The *Lancet*, 1881, vol. ii., p. 430, under the heading of "The Burial of Still-born Infants," states that "Greater security for the due observance of these necessary regulations (the Births and Deaths Registration Act of 1874), for the burial of infants said to be still-born, is urgently called for. It is constantly patent that the burial of deceased infants as still-born, if checked, is by no means prevented; and that the authorities of burial-grounds, by their laxity in carrying out the provisions of the Act, afford dangerous facilities for the concealment of crime, or negligence, and for a practice which threatens to impair the value of our birth and death registration statistics; for, if a live-born infant be buried as still-born, neither its birth nor its death is registered."

A case of forceps-delivery occurred in the hands of the writer (E. P. V.), in which the child, when extracted, was quite purple in colour, and absolutely dead to all appearances—there was no breathing nor impulse to be found anywhere. After some efforts at resuscitation in the way of artificial respiration—not very thoroughly done, nor much prolonged (for the child was believed to be dead)—with a warm bath and frictions, it was laid aside and covered up. At a subsequent visit some hours later, the child was found in the nurse's lap, completely recovered, and changed in colour to a bright pink. The nurse said she did not like to give the little fellow up, and by breathing into his mouth for some time he showed returning life, and by keeping it up he soon began to breathe himself.

Cases like this are believed to be not infrequent, because physicians and nurses are not, as a general rule, aware of the great tenacity of life possessed by the new-born infant.

"*Still-births* are not registered in England; but, under the new Registration Act, no still-born child can be buried without a certificate from a registered practitioner in attendance, or a declaration from a mid-wife, to the effect that the child was still-born. The proportion of still-births in this country is supposed to be about four per cent., but this is uncertain."—A. NEWSHOLME, *Vital Statistics*, 1889, p. 61.

"The proportion of deaths from premature births, compared with the total number of births, in 1861-65 was 11·19 to 1,000 births; since which time it has steadily increased, reaching the ratio of 15·89 per births in 1,000 in 1887."—*Ibid., p. 216.*

The same author, p. 17, states that "a certain proportion of the *births remain unregistered* (a). There is strong reason for thinking that a certain number of children born alive are buried as still-born."

APPENDIX C.

RECOVERY OF THE DROWNED.

THIS is perhaps the best known and most generally appreciated occasion of rescuing the apparently dead. The high degree in which it has excited public sympathy will appear from a glance at that section of the "Bibliography" (towards the end of the eighteenth century) which gives the titles of essays and reports connected with the Royal Humane Society and the corresponding foreign institutions upon which our own was modelled. The following general remarks and cases are from the essay of Dr. Struve, of Görlitz, Lusatia, 1802 :—

"A great number of persons apparently drowned have been restored to life without the use of stimulants, merely by the renovated susceptibility of irritation. I have collected thirty-six cases of persons apparently drowned in Lusatia from the year 1772 to the year 1792. Most of them were treated by uninformed people, and revived by friction and warming ; two persons, however, were indebted for their lives to the continuation of the resuscitative process for several hours. The greatest number were children ; which is to be ascribed not only to the greater danger to which they are exposed of drowning, but also to the longer continuance of vital power in the infant frame" (p. 136).

"A boy of about a year and a half old had lain upwards of a quarter of an hour in the water, and was found face downwards, and the whole body livid and swollen. He was undressed, wiped dry, and wrapped in warm blankets : but the most particular part of the process was rolling the body upon a table, shaking it by the shoulders, and rubbing the feet. This having been continued for an hour, a convulsive motion was observed in the toes ; sneezing was excited by snuff ; the tongue stimulated by strong vinegar : the throat irritated with a feather ; an injection given. The child vomited a large quantity of water, and in an hour afterwards began to breathe, and was completely restored to life." (p. 137).

"A woman upwards of thirty years of age, and who was affected with epilepsy, fell in a fit from a height of twenty feet into the water, where she remained a full quarter of an hour before she was taken out. Mr. Redlich, surgeon, of Hamburg, had her put into a bed warmed by hot bottles ; she was rubbed with warm flannels, some spirits were dropped into her mouth, when in a quarter of an hour symptoms of life, such as convulsive motion and a very weak pulse, appeared. In three hours from the time she was taken out of the water she recovered completely" (p. 138).

Dr. Charles Londe, in a remarkable pamphlet ("Lettre sur la Mort

Apparente, les Conséquences Réelles des Inhumations Précipitées, et le Temps Pendent lequel peut persister l'Aptitude à étre Rapellé à la Vie." Paris, Bailliére, 1854), records some instances of narrow escapes from premature burial of the drowned, one of which may be cited :—

"On the 13th of July, 1829, about two p.m., near the Pont des Arts, Paris, a body, which appeared lifeless, was taken out of the river. It was that of a young man, twenty years of age, dark-complexioned, and strongly built. The corpse was discoloured and cold; the face and lips swollen and tinged with blue; a thick and yellowish froth exuded from the mouth; the eyes were open, fixed, and motionless; the limbs limp and drooping. *No pulsation of the heart nor trace of respiration was perceptible.* The body had remained under water for a considerable time; the search for it, made in Dr. Bourgeois's presence, lasted fully twenty minutes. That gentleman did not hesitate to incur the derision of the lookers-on by proceeding to attempt the resuscitation of what, in their eyes, was a mere lump of clay. Nevertheless, several hours afterwards, the supposed corpse was restored to life, thanks to the obstinate perseverance of the doctor, who, although a strong man and enjoying robust health, was several times on the point of losing courage and abandoning the patient in despair. But what would have happened if Dr. Bourgeois, instead of persistently remaining stooping over the inanimate body, with watchful eye and *attentive ear*, to catch the first rustling of the heart, had left the drowned man, after half an hour's fruitless endeavour, as often happens? The unfortunate man would have been laid in the grave, *although capable of restoration to life!*"

To this case, Dr. Bourgeois, in the "Archives de Medecine," adds others, in which individuals remained under water as long as SIX HOURS, and were recalled to life by efforts which a weaker conviction than his own would have refrained from making. These facts lead Dr. Londe to the conclusion that, *every day, drowned individuals are buried, who, with greater perseverance, might be restored to life!*

The following case in point appears in the *Sunnyside*, New York, communicated by J. W. Green, M.D. :—

"A few years since I was walking by the Central Park, near One Hundred and Tenth Street and Fifth Avenue. Noticing a crowd that was acting in an unusual manner by the side of the lake, I approached and inquired of one of the bystanders what was the cause of the excitement. He replied, 'A boy is drowned.' I advanced to the edge of the water, and saw two or three men in the water searching for the body. As they had not yet discovered it, I made enquiries, and found at last a small boy who had been a comrade of the victim. He showed me the spot from which the boy had fallen. I then pointed out to the searchers where to look, and immediately the body was recovered. I took it at once from the hands of the person who had it, and held it reversed, in order to disembarrass it of all the water possible, for a minute or two, then stripped it of its clothing, sent for a blanket and brandy. I took a woollen coat from one of the bystanders until the blanket should arrive, laid the child upon it and commenced to rotate it. This I continued to do for at least fifteen minutes by the watch. I then tried auscultation ; no murmur could be heard.

"The skin was cold, the lips were blue. Every artery was still. With all these signs of death present it was still obligatory upon me to persevere. At the end of fifteen minutes there was a slight gasp. A small quantity of brandy was placed upon the tongue. A little of this ran into the larynx, and the stimulation was sufficient to produce a long inspiration and then a cough. This was more than a half-hour from the time when the boy had been removed from the water. Complete restoration did not occur until nearly an hour from that time. He was now given to his mother, and I was informed on the following day that he entirely recovered, without an unfavourable symptom."

The three following cases of resuscitation from apparent death by drowning are copied from the most recent reports of the Royal Humane Society, London :—

"On 13th of August, 1895, Samuel Lawrence, aged five years, while playing on the bank of a disused clay-pit at South Bank, Yorkshire, fell into the water and sank. Two of his companions dived into the water, and brought him up after a submersion of from seven to ten minutes in an unconscious state. Two working men commenced artificial respiration, and Dr. Steele continued it for ten hours before the boy showed signs of returning sensibility and his complete recovery."

"October 6th, 1895.—At Deptford, Surrey, a woman with a baby in her arms threw herself into the canal. They were rescued by the Royal Humane Society's drags. Two ladies took possession of the bodies (time of submersion not stated), and they employed Silvester's system of artificial respiration with success, in the case of the woman in about one hour, and with the child one hour and a half."

"August 6th, 1895.—At Bradford, England, Rudolf Pratt, a clerk with Midland R.R. Company, was bathing, and sank in deep water. A bystander by diving brought him up. After a submersion of five minutes, unconscious, and not breathing, Dr. Oldham restored respiration by Sylvester's method after one and a half hour's treatment."

These three cases are instructive on account of the length of time animation remained suspended before it could be aroused to a state of activity ; and they lead to the belief that many cases that are given up as actually dead could be saved if efforts at resuscitation were kept up for a lengthened period, as in the first case.

In cases of drowning some persons are quickly revived after a long submersion; others again who are under water only a short time require artificial respiration for a long time before they show signs of returning life, as was the case with Samuel Lawrence, who was submerged only ten minutes, yet required ten hours' active treatment to revive him.

APPENDIX D.

MISCELLANEOUS ADDENDA.

HASTY BURIALS.

As an illustration of hasty burials dealt with in Chapter X. the following case is cited from the *King's County Chronicle*, Parsonstown, Ireland, August 27, 1896:—

"ROSCREA GUARDIANS.

"Thursday—Present: T. Jackson, D.V.C., in the chair; L. S. Maher, J.P.; M. Bergin, J.P.; W. J. Menton, W. Jackson, P. Roe.

"Mr. Roe—You made short work of Jack Ryan at the chapel of Knock. He was alive and speaking at three o'clock, and buried at six the same day. The Master stated that, it being supposed the man died from an infectious disease, no person would assist in coffining him till a message came asking that he (the Master) would send out some of the male inmates, and he sent two and had him coffined and interred. Mr. Roe—The man was not cold when he was buried. Master—The nun tells me the man had an ounce of tobacco clasped tightly in his hands. Chairman—What disease had he? Clerk—Pneumonia was certified by the doctor. The people believed that he had died from an infectious disease, and insisted he should be buried immediately. Mr. Roe—It was certainly short work—a man dying at three o'clock and buried at six. Master—This man was married to a woman who was a nurse in the old Donoughmore workhouse, and they lived at Drumar, Knock."[1]

---

[1] With reference to the burial customs in Ireland, the *King's County Chronicle*, Parsonstown, September 17, 1896, says:—"Young children are buried the day after death, but adults are waked for two, and sometimes three, nights."

EVIDENCE OF RESUSCITATIONS IN GRAVE-YARDS.

Reference has been made in this volume to the discoveries of premature burial brought to light during the investigations of charnel-houses in France, and the removal of grave-yards, necessitated through the rapid expansion of towns, in America. The *Casket*, Rochester, New York, U.S., of March 2, 1896, gives a detailed narrative of recent discoveries made by T. M. Montgomery in the removal of Fort Randall Cemetery, with the condition of the bodies found as to decay or state of preservation, and says :—

" We found among these remains two that bore every evidence of having been buried alive. The first case was that of a soldier that had been struck by lightning. Upon opening the lid of the coffin we found that the legs and arms had drawn up as far as the confines of the coffin would permit. The other was a case of death resulting from alcoholism. The body was slightly turned, the legs were drawn up a trifle, and the hands were clutching the clothing. In the coffin was found a large whisky flask, showing that those who buried him were not his friends, or else that they too were afflicted with the disease that had cut short the life of their companion.

" It occurred to us at that time that this was a great argument in favour of incineration. Nearly two per cent. of those exhumed here were, no doubt, victims of suspended animation. Once before in our experience have we noted this; and, while not believing in as large a percentage of live burials as the radical advocates of cremation claim, yet we know that the percentage is larger than most scientists give. Disinterment is the only solution of the question. In regard to these two cases, we wish to say that science has proved that electricity does not always kill, and that persons addicted to the liquor habit, after long debauches, sometimes relapse into a comatose state, and are to all appearances dead. Statistics show that a great many die annually of these causes, hence the percentage in cases of this kind must be very large. What is the remedy ?"

HASTY EMBALMMENTS IN THE UNITED STATES.

The *Casket*, Rochester, New York, September, 1896, observes :—At different times considerable opposition has been raised against embalming by Boards of Health and other officials in various localities, on account of the haste with which the embalmer proceeds with his duties. A few recent cases of supposed corpses recovering, one of which occurred in Philadelphia, Pa., have revived the question, and it is reported that the Philadelphia Board of Health may take action looking to the enactment of a law prescribing the period of time which should elapse after death before a body should be embalmed.

In a recent issue of the Philadelphia *Times*, Funeral-Director John J. O'Rourke, a well-known professional of that city, expresses himself on the subject as follows :—

" These two narrow escapes from burial alive have further impressed me with one of the perils attending the disposition of the dead—I mean the danger of hasty embalming. As you know, in most cases the doctor who has had the patient is not called in after death, and very often the relatives of the deceased expect the undertaker, if embalming is to be done, to proceed with it at once. All the embalming schools teach that the only proper way to thus treat the body is by use of fluids through the arteries. But in the lectures on the subject no period that should be permitted to elapse before it is begun is prescribed, and, as a rule, it follows dissolution as quickly as possible.

" I contend that there should be some law or official rule governing the matter, because after the artery is punctured and the fluid goes through the whole body, it is sure to destroy any spark of life that might remain. I have never met with any cases of resuscitation myself, but have had instances of deaths that made me hesitate in the work of embalming. Some months ago a man came to me fifteen minutes after a relative had breathed his last, and asked me to embalm the body. I went to the house, and, after seeing the corpse, refused, saying that I would not do it until after the expiration of twelve hours. The man had died of consumption, yet, for fear of it being a case of suspended animation, I would take no chances.

" At another time a person had died of dropsy. Within half an hour I was summoned. The attending physician had not been there, and twenty-four hours afterwards he gave a certificate of death from cancer. The body was very warm when I arrived, and neighbours who had kindly volunteered to prepare it were doubtful if life was extinct. I had the corpse laid on an embalming table for two hours, and then placed it in what is known as a Saratoga patent box, in which are pans filled with salted ice, so arranged that cold air circulates around the body. Had this been a case of suspended animation, it would have taken several hours to dispel the heat within the corpse.

" Of course there are some supposed unmistakable signs. The only positive signs of dissolution are those which depend on molecular change or death-rigidity of the muscles of the whole body, and putrefaction of the tissues. These are most marked in organs and tissues the vital functions of which are the most active. The action of the heart, the movements of respiration, may be reduced as to be altogether imperceptible, so that the functions of circulation and respiration appear to be arrested. This is occasionally observed in temporary syncope, in which a person to all appearances dead has, after a time, regained consciousness and recovered.

"The peculiar condition of the nervous system called catalepsy, and the state of trance, are likewise further examples of the so-called apparent deaths; but, on the occurrence of actual death, the irritability of the muscles by degrees disappears, electricity no longer excites their contraction, and then cadaverous rigidity sets in. . . . Some action will, in all probability, be urged upon the next Legislature or upon the Board of Health."

APPENDIX E.

SUMMARY OF ORDINANCES, ETC., RELATING TO THE INSPECTION OF CORPSES AND OF INTERMENTS.

IN the sixteenth Council of Milan, Saint Charles Borromeo prohibited burials before twelve hours after ordinary cases of death, and twenty-four hours after cases of sudden death. As early as the sixteenth century serious attention in the examination of the dead was made obligatory by the enactment of Article 149 of the Criminal Statutes of Charles the Fifth. This was the foundation of legal medicine in Germany. In France, a similar ordinance was first established in 1789.

NETHERLANDS.

Act of April 10th, 1869.

No burial is allowed without the written permission of the Civil Recorder, granted upon the production of a certificate of a qualified physician, and not until thirty-six hours have elapsed after death, nor later than the fifth day after death. But this regulation can be set aside, and a longer period allowed, by the Burgomaster, on the application of a doctor.

Dead-houses are in use for bodies dead of infectious diseases.

FRANKFORT-ON-THE-MAIN.

Death must first be established by a licensed physician, who carefully examines the body for that purpose, and, if satisfied, then issues a certificate which states the name, age, sex, place, and date, and immediate cause of death. The certificate is taken within twenty-four hours after the death to the Standesamt, where the death is recorded, and a certificate to that effect is given, and presented to the Cemetery Commission, which assigns the place of burial.   The corpse is required to remain unburied three days, either at the place of death or at the mortuary, where it is under the observation of attendants; but there is no State-appointed inspector of the dead, nor electric bells or other means for announcing and recording any movements of the body.   The system of inspection and certification by qualified physicians, with the delay of three days, and the favourable condition of the dead-houses, have been the means of preventing the living from being mistaken for the dead in a number of cases.

FRANCE.

Interments must not take place, according to Article 77 of the Code Napoleon, before twenty-four hours of death, but in practice it is twenty-four hours after death-notification by the *mort - verificateur*.   During epidemics, or when deaths occur from infectious or contagious diseases, the interments must invariably be made within twenty-four hours of death.

Article 77 of the Civil Code states that "No burial shall take place without an authorisation, on free paper and without expense, of the officer of the Civil State, who will not be empowered to deliver it, unless after having visited the deceased person, nor unless twenty-four hours after the decease, except in cases provided for by the regulations of the police." It results from this that no corpse can be buried before a minimum delay of twenty-four hours shall have expired after the decease.   The formal record of the decease must be made by the officer of the Civil State (the mayor), or, which is what takes place in most of the communes, by a medical man delegated by the mayor, and who takes the title of medical officer of the Civil State.

The Article 77 of the Civil Code is generally strictly observed in Paris and in other cities of France.   The obligation to await the delay of twenty-four hours is intended to prevent too hasty burials.   One considers, in fact, that that delay is generally necessary in order to be able to have certain proofs of death.

By Article 358 of the Penal Code, the burial of a deceased person without such authorisation is punishable by a maximum period of two months' imprisonment, and a maximum fine of fifty francs, without prejudice to other criminal proceedings which may be applicable under the circumstances.

Exceptions, however, have been established in certain cases.   For example, in times of epidemics, or of too rapid decomposition of the corpse in the usual case, there is urgent need, in fact, to bury the body of a person attacked with a contagious or epidemic malady, in order to suppress one of the causes of propagation of the epidemic, or of the contagion. In the second case, it is understood that one could not keep longer, without danger to the public health, a corpse in complete putrefaction. There is occasion also to observe that, in these circumstances, the end which the legislator has proposed to himself is equally obtained, since there cannot be any doubt as to the real death. However that may be, it is the mayor (officer of the Civil State) to whom it appertains, according to the terms of the Article 77 of the Civil Code, to give authority to bury; and if he gives that authorisation before the expiration of the delay of twenty-four hours, it is after having established by himself, or by the medical officer of the Civil State, the fact of its necessity, resulting from the circumstances of which we have just spoken.

It is to be remarked that the Article 77 fixes a *minimum* and *not a maximum* delay. It is always the mayor to whom it appertains to fix the day and the hour of the burial, and there may happen such and such a circumstance which necessitates a delay of the obsequies. The mayor need only assure himself in that case that no danger will result to the public health, which naturally is the case when the corpse is embalmed, or is placed in a leaden coffin.

Outside Paris and other large cities, and especially in the rural districts, much laxity prevails both as to verification of death and the time of burial, and cases of premature burial are not infrequent.

AUSTRIA.

The laws relative to funerals and burials are very strict—perhaps the most thorough in their requirements of any in Europe. They provide for a very careful inspection of the body by medical inspectors, quite independently of the attending physicians, in order to ascertain if the death be absolute. Minute and specific official directions guide them as to the method of examination and the signs of death to be looked for. And they further provide for carrying out any particular method, as to which the deceased may have given directions, in order to prevent a possible revival in the coffin. Should the surviving relatives desire it, a *post-mortem* operation may be made upon the body, in the presence of the medical inspectors and the police; in which case the heart is pierced through; and a full report of the operations must be forwarded to the civic magistrate. A fee of six florins is allowed for such an operation.

CITY OF VIENNA.

Every death to be inquired into by the municipal physician. The first of five objects is to ascertain whether the person be really dead. In examining whether there are any remaining indications of life, he will rely not upon any one sign, nor even upon putrefaction, but upon the totality of the signs of death. If there are any indications of life remaining, he must at once institute the means of resuscitation approved by science, and continue them until such time as the family medical attendant is assured of their uselessness. If there be any doubt as to the reality of the death, a second inspection of the body is to be made by the municipal physician within twenty-four hours. Burial, as a rule, is not to be until forty-eight hours after death; but the interval may be shortened in cases of infectious diseases or of unusually rapid decomposition.

PROVINCE OF DALMATIA.
Vice-Governor's Order of 29th April, 1894.

Every death to be inquired into by the parish physician, or a deputy appointed by the mayor. The first of six objects of the inquest is to ascertain whether the person be really dead. In the event of a non-medical examiner discovering signs of life, he is to send for a doctor. Inasmuch as decomposition, the only sure sign of death, is, as a rule, a phenomenon of later occurrence than the time appointed for the inquest (within twelve hours of the notification of death), the examining person must base his certainty of the extinction of life, not upon one sign, but upon the totality of the signs of death.

KINGDOM OF SAXONY.
Law of 20th July, 1850.

The burial of a corpse must not take place until seventy-two hours after death, and the signs of decomposition are clearly visible. Any proposed departure from this rule, in the event of earlier putrefaction, or the absence of decomposition at the end of seventy-two hours, requires the authority of a physician called in. By the above Law, the following Orders are suspended: (1) the Order of 11th February, 1792, concerning the treatment of the dead, and the precautions necessary to prevent the apparently dead from being buried prematurely; (2) the General Order of 13th February, 1801, concerning precautionary measures in the burial of those dead of infectious diseases; (3) the Law of 22nd June, 1841, together with the Administrative Orders, concerning the examination of corpses and the establishment of mortuaries.

CITY OF MUNICH.
Order of 30th October, 1848.

The ordinance hitherto in force, as to making an incision in the sole of the foot in cases of patients who die in the hospitals, is abolished; the hospital physicians to use their discretion whether or not the incision should be made; but, in cases for which is demanded an earlier burial than is usually prescribed, whether they have been hospital or private patients, the incision is to be made in the sole of the foot at the end of the second inspection, and every other means taken to ascertain whether the death be apparent or real.

CALCUTTA.

1. The prevailing custom for Christians and Mahomedans is to bury the dead. The Hindoos burn them as a rule, but many prefer to throw them into a sacred river, particularly the Ganges or its tributaries, if they can do so unmolested by the authorities.

2. There are no mortuaries. The signs which are assumed to indicate death are the various conditions and appearances when animation is suspended.

3. Cases of revival from supposed death are sometimes heard of among the Hindoos, who regard such persons as outcasts. If the signs of returning life are not very manifest when a person begins to revive, he is sometimes killed by stuffing the mouth and nose with mud, which generally accomplishes the object.

BOMBAY.

1. There are no laws or regulations in India for the disposal of the dead. The customs and formalities follow the traditions and requirements of religious belief.

  a. The Hindoos burn their dead immediately after death takes place.

  b. The Parsees take their dead to a "Tower of Silence" as soon as death takes place, and, after certain prescribed ceremonies, the body is speedily devoured by vultures.

  c. The Europeans and Mahomedans bury their dead within from twenty-four to forty-eight hours, because putrefaction usually sets in soon after death on account of the heat and humidity of the climate.

2. There are no mortuaries, excepting in connection with hospitals, where observations can be made.

CAPE TOWN, AFRICA.

1. There are no laws nor regulations relative to the disposal of the dead, excepting in cases requiring an inquest or *post-mortem* examination. The custom is to bury within twenty-four to thirty hours after death, but the time is sometimes extended to two or three days.

2. There are no dead-houses, except at the hospitals, which are under the management of the superintendent.

24

3. The certificate of the medical attendant is sufficient for burial purposes. The complete cessation of respiration and the heart's action are considered an absolute indication of death. When decomposition sets in, it usually appears within twenty-four hours after death, although in winter that process may be longer delayed.

MOSCOW.

Orthodox Russians keep their dead three days before burial. During that time the body lies with the face uncovered, and a deacon chants and prays over it twice a day. A medical certificate of death is imperative before burial.

BRUSSELS.

Burials are regulated by the Communal Council in accordance with law. The system is complicated, but thorough. The medical men connected with the Government Medical Service ("Doctors of the Civil Government") have the sole control of the examinations of deaths, as well as births, accidents, sudden deaths, suicides; and attend to burials, autopsies, postponements of burials, etc., on their own motion. Interments usually take place within forty-eight hours of death, but they may be carried out sooner during epidemics for the public safety.

There are mortuaries in the city and suburbs, to which bodies may be taken at the request of surviving relatives, or by the order of the health authorities, according to private necessities or for the public safety. Except by the special authorisation of the officers of the civil government, bodies cannot remain in the mortuaries longer than forty-eight hours; and a burial cannot take place in less than twenty-four hours. Special care is taken to test the reality of death in still-born infants, and efforts are made to revive them, as well as all other cases of seeming death. In cases of women dying during advanced pregnancy, the infant must be roused by artificial respiration, in order to restore animation if possible. The process for obtaining a delay for burial is intricate and cumbersome, and to a foreigner unaccustomed to the language and the local usages the chances would be against securing such a permit before the time allowed for burial had transpired.

DENMARK.

Mortuaries are connected with all the churches, cemeteries, and some of the hospitals, and are growing in favour in the country places; but as yet they are unprovided with any appliances for the resuscitation of the apparently dead, or for the prevention of premature burials. No corpse, however, is allowed to be taken to a mortuary before it has been inspected, and a death-certificate issued by a qualified physician; but, when this is done, death is considered absolute. No corpse is allowed to remain in any church, chapel, or mortuary longer than seven days after supposed death, without special permission. Coffins that contain bodies which have died from infectious diseases must be so indicated, and cannot be opened in the mortuaries.

As a rule, bodies are kept seventy-two hours before burial. The signs that are considered sufficient to establish death are the glazed appearance of the eyes, livid spots on the skin, and muscular rigidity. In doubtful cases, the time before burial can be extended by authority of the Board of Health, of which the Police Director is a member.

SPAIN.

Burials usually do not take place until twenty-four hours after death. For example, if a death takes place about four p.m., the burial is made late in the following afternoon. In time of epidemic, bodies are hurried to the cemeteries, where depositories are provided, which are under the care of watchers until the expiration of twenty-four hours after death. The certificate of a reputable physician as to death is sufficient to authorise burial. Relatives or friends usually remain with the body until burial, excepting in cases when judicial proceedings are held over it to determine the circumstances of the death.

IRELAND.

There are no laws in Ireland regarding the disposal of the dead, but the Sanitary Acts of the United Kingdom can be applied in any case within a reasonable period, on the ground of public health. There is no fixed period for keeping a body before burial. The Roman Catholics usually bury on the third or fourth day after death; but in some districts custom sanctions burial within twenty-four or thirty-six hours. Local burial authorities sometimes require a medical certificate before burial, but, there being no legal obligation for it, this is often omitted. In cases of suicide, sudden death, or death by violence, the Coroner holds an inquest, and gives a certificate accordingly.

There are no dead-houses in Ireland, where bodies may be observed for a period of time before burial.

Concerning burials in England, see Glen's " Burial Acts" for the general burial practice; also " Regulations for Wilton Cemetery."

THE UNITED STATES.

In the United States of America, as a rule, everything relative to the disposal of the dead is regulated by local Boards of Health, as authorised by State laws. A burial cannot take place without a certificate from a legally licensed physician, which must state the cause of death; the place and time when it occurred: the full name; age; sex; colour; occupation; birth-place; names and birth-places of both parents. There are no laws or regulations that require the inspection of the body to verify the fact of death (the certificate, as in England, as to the cause is considered sufficient for this purpose), and no time is fixed when a body must, or must not, be buried. This is regulated by, and left to, the convenience of the family of the deceased, by the season of the year, by the opinion of the attending physician, etc. But the Health Officers can order the burial whenever, in their opinion, the public health requires it. As a rule, burials after supposed death are made sooner in the South, and among the poor, than in the North, and among the well-to-do classes. In remote unsettled regions burials not seldom take place without these formalities, and they are often carried out in a hasty manner; but usually they do not take place till three days after supposed death, and sometimes, particularly in cold weather, a longer time is allowed. All large cemeteries have chambers for the temporary deposit of bodies, but they are not under observation, as it is taken for granted that they are dead.

APPENDIX F.

THE JEWISH PRACTICE OF EARLY BURIAL.

R. J. WUNDERBAR, in his standard work on "Biblisch-talmudische Medicin," Riga and Leipzig, 1850-60, gives, in pp. 5-15 of the concluding section (Abtheil. 4, Bd. ii.), the following summary of the origin of the peculiar Jewish practice of burying the corpse within a few hours of death:—

In the Levitical law (Num. xix. 11-22) every dead body was an unclean thing, including those dead in the tent and on the battlefield. Touching a corpse involved purification and separation for seven days. This ordinance is supposed to have had a sanitary motive, having probably originated with cases of infectious disease. There is only one Biblical ordinance as to early burial, and that is indubitably restricted to persons executed for crime: Deut. xxi. 22, 23, "And if a man have committed a sin worthy of death, and he be put to death, and thou hang him on a tree, his body shall not remain all night upon the tree, but thou shalt in any wise bury him that day (for he that is hanged is accursed of God), that thy land be not defiled which the Lord thy God giveth thee for an inheritance."

This statutory limit to the exposure of the bodies of malefactors was the most convenient way of checking the practice, common in other countries, of leaving corpses of criminals to hang upon the gibbet until they rotted, or were consumed by birds of prey. Its motive was to prevent, by the promptest measure, an indefinite degree of neglect in altogether special cases.

There is nothing else in the Bible concerning early burial; on the contrary, the patriarchal practice, in the case of eminent persons, seems to have been to keep the body for a considerable time above ground, after the manner of Egypt. Prior to the Babylonian exile there is not a trace of the later practice of speedy burial. The post-Talmudic custom had arisen entirely from a misunderstanding. It is true that the Talmud enjoins that corpses—according to circumstances—be kept unburied not longer than one day; but it also permits them to lie above ground for days, so that elaborate funeral preparations might be made, or time given for mourners to arrive from a distance. Lastly, the Talmud relates the burial of one apparently dead who revived and lived for twenty-five years, and begat five children; whereupon a rabbinical ordinance was made that the corpse (which would have been laid in a vault or in a tomb above ground) should be visited diligently until three days after death. (The references to the Talmud are: Semachoth 8: Moedkaton 1, 6; Sabbat 151, 152: Sanhedrin 46a.)

Wunderbar admits that there had been cases of premature burial among the Jews, but he asserts their extreme rarity, and doubts the authenticity of most of the traditional or historical cases in general.

In Jewish circles in Germany towards the end of last century there was much controversy as to the inexpediency of the practice of early burial. In the " Berlinische Monatschrift " for April, 1787, p. 329, (cited by Marcus Herz, " Ueber 'die frühe Beerdigung der Juden," Berlin, 1788. p. 6,) there is printed a letter from Moses Mendelssohn to the Jews of Mecklenburg, in which he advises them to keep their dead unburied for three days. " I know well," he adds, " that you will not follow my advice ; for the might of custom is great. Nay, I shall perhaps appear to you as a heretic on account of my counsel. All the same, I have freed my conscience from guilt."

The above-cited essay by Dr. Marcus Herz, of Berlin, arguing against the Jewish practice, called forth a reply by Dr. Marx, of Hanover, who was of opinion that the burial might safely proceed after the body had been left on the bed for three hours, and had then been pronounced lifeless by the medical attendant, according to the practice in that part of the country. To that Dr. Herz rejoined, in a second edition, that the medical attendant was no better judge than an ordinary man, inasmuch as all experimental tests were fallacious, and decomposition the only sure sign. He cites the following statement by an experienced Jewish physician, Dr. Hirschberg, of Königsberg (from the Jewish periodical, "Sammler," vol. ii., p. 153) : — "I have practised medicine for forty years, and have always grieved over the practice amongst us of too hasty burial of the dead—on the day of decease. It happened once in my practice that a woman lay for dead three days and then awoke and revived. At first I would not allow the body to be moved from the bed, but the undertaker's men violently resisted me, taking up the body and laying it on the ground. According to their custom, they would have buried it the same day, had I not earnestly called out to them : ' Beware lest you do lay her in the ground this day ! She is still alive, and the blame will be on you.' I had her covered with warm, woollen clothes; on the following morning some signs of life were manifest; she lay still, and gradually awoke out of her death-slumber."

Herz declared, as Wunderbar did subsequently, that the passages in the Talmud on which the Jewish custom was based had been misinterpreted ; and he specially accused the rabbis Jacob Emden, of Altona, and Ezechel, of Prague, of rabbinical subtilty on the one hand, and of a fallacious dependence upon scientific signs of death on the other.

———— ————

At the World's Medical Congress (Division of Eclectic Medicine), held in Chicago, June 3, 1893, the following resolution was proposed by Dr. John V. Stevens, and adopted :—

"Whereas we believe that many persons in the past, in the condition simulating death from various causes, have been buried alive ; therefore,

"Resolved—That it should be the duty of all Governments to pass laws prohibiting the burial of bodies without positive proofs of death ; that the nature of these proofs should be taught in all schools and printed in all newspapers throughout the world."

BIBLIOGRAPHY.

SEVENTEENTH CENTURY.

KORNMANNUS (Henricus). De miraculis mortuorum. Francof., 1610.

TIRELLUS (Mauritius). De causis mortis repentinae. Venet.. 1615.

ZACCHIAS (Paulus). Quaestiones medico - legales. Lib iv. cap. i.,
quaest. xi , " De mortuorum resurrectione," fol. 241-247 of editio
tertia. Amstelaedami, 1651.

[Gives many of the classical cases, with critical remarks.]

KIRCHMAIER (Theodor) and NOTTNAGEL (Christoph). Elegantissimum
ex physicis thema de hominibus apparenter mortuis. Wittenbergae,
1670.

[Collects cases, from ancient and more recent writers, of the apparently dead having been taken for dead:—Pliny, Hist. Nat., lib. vii. 52 ;
Plutarch, De sera numinis vindicta ; Apuleius, Floridorum, lib. vi. ;
St. Augustine, De cura mortuorum ; Thuanus (no ref.) ; Diomed
Cornarus, Hist. admirand. (case of a Madrid lady who is supposed to
have given birth to a child after she was laid in the tomb, the corpse
having a new-born dead infant in the right hand when the vault was
opened a few months after) ; Chr. Landinus, notes to Virgil, Æn. vi.
(incident at a funeral, of which he was an eye-witness at Florence) ;
Horst. Med. mir., cap. ix. (woman left for dead of the plague at
Cologne in 1357) ; and the case of a glazier, then living at Wittenberg.
who was treated as dead when a child of three years.]

GARMANN (L. Christ. Frid.). De miraculis mortuorum libri tres, quibus
praemissa dissertatio de cadavere et miraculis in genere. Opus
physico-medicum curiosis observationibus experimentis aliisque rebus
exornatum. Ed. L. J. H. Garmann. Dresden and Leipzig, 1709.
(First ed., Leipzig, 1670.)

BEBEL (Balthasar). Dissertatio de bis mortuis. Jena, 1672.

EIGHTEENTH CENTURY.

HAWES (Dr.). On the duty of the relations of those who are in dangerous illness, and the hazard of hasty interment. A sermon preached in the Presbyterian Chapel of Lancaster in 1703, wherein it is clearly proved, from the attestation of unexceptionable witnesses, that many persons have been buried alive.

LANCISI (Johannes M.). De subitaneis mortibus libri duo. Romae, 1707; Lucae, 1707; Lipsiae, 1709.

WILFROTH (Johannes Christianus). Dissertatio de resuscitatione semi-mortuorum medica. Halae, 1725.

RANFT (Michael). Tractat von den Kauen und Schmatzen der Todten in Gräbern, worin die wahre Beschaffenheit derer Hungarischen Vampyrs gezeight, etc. Leipzig, 1734.

BEYSCHLAG (Fr. Jac.). Sylloge variorum opusculorum. " De hominum a morte resuscitatorum exemplis." Halae Sueviorum, 1727-31.

WINSLOW (Jacques Benigne), Professor of Anatomy at Paris. An mortis incertæ signa minus incerta a chirurgicis quam ab aliis experimentis. Paris, 1740. Dissertation.

———— Dissertation sur l' incertitude des signes de la mort, et l' abus des enterremens et embaumemens precipités; traduite et commentée par Jacques Jean Bruhier. Paris, 1742. (With the Latin text.)

BRUHIER (Jacques Jean), d'Ablaincourt. Mémoire sur la necessité d' un règlement général au sujet des enterremens et embaumemens—addition au mémoire presente au Roi. Paris, 1745-46.

———— Dissertation sur l' incertitude des signes de la mort, et l'abus des enterrements et embaumemens precipités. Second ed. Two vols. Paris, 1749.

———— The uncertainty of the signs of death and the danger of precipitate interments and dissections. Second ed. London, 1751.

[Bruhier, in his work Dissertations sur l' incertitude des signes de la mort et l'abus des enterremens, produces accounts of one hundred and eighty-one cases, among which there are those of fifty-two persons buried alive, four dissected alive, fifty-three that awoke in their coffins before being buried, and seventy-two other cases of apparent death.]

ANON. The uncertainty of the signs of death, and the danger of precipitate interments and dissections demonstrated. Dublin, 1748.

COOPER (M). Uncertainty of the signs of death, precipitate interment and dissection, and funeral solemnities. London, 1746.

JANKE (J. G.). Abhandlung von der Ungewissheit der Kennzeichen des Todes. Leipzig, 1749.

LOUIS (Antoine). Six lettres sur la certitude des signes de la mort, où l'on rassure les citoyens de la crainte d'etre enterrés vivans ; avec des observations et des experiences sur les noyés. Paris, 1752.

PLAZ (Antonius Gulielmus). De signis mortis non solute explorandis. Specimen primum, Lipsiae, 1765 ; secundum, 1766 ; tertium, 1766 ; quartum, 1767.

———— De mortuis curandis. Diss. Lipsiae, 1770.

MENGHIN (Joh. Mich. de). Diss. de incertitudine signorum vitae et mortis. Vienna, 1768.

ESCHENBACH (Christ. Ehrenfr.). De apparenter mortuis. Vienna, 1768.

JANIN DE COMBE BLANCHE (Jean). Reflexions sur le triste sort de personnes qui sous un apparance de mort ont été enterrées vivants, etc. Paris, 1774.

DE GARDANE (Joseph Jacques). Avis au peuple sur les asphyxies ou morts apparentes et subites. Paris, 1774. Portuguese transl. included in Avisos interessantes sobre as mortes apparentes. Lisbon, 1790.

———— Catechisme sur les morts apparentes. dites asphyxies. etc. Paris, 1781.

NAVIER (Pierre Toussaint). Réflexions sur les dangers des inhumations precipitées et sur les abus des inhumations dans les eglises, etc. Paris, 1775.

PINEAU (————). Mémoire sur le danger des inhumations precipitées, et sur la necessité d'un règlement pour mettre les citoyens à l'abri du malheur d'etre enterées vivans. Niort, 1776.

MARET (Hugues). Mémoire pour rappeler à la vie les personnes en état de mort apparente. Dijon, 1776.

BRINKMANN (Joh. Pet.). Beweis der Möglichkeit dass einige Leute können lebendig begraben werden, etc. Düsseldorf, 1777.

SWIETEN (Baron Geerard Van). De morte dubia. Vienna, 1778.

TESTA (Antonio Guiseppe). Della morte apparente. Firenze. 1780.

DOPPET (F. A.). Des moyens de rappeler à la vie les personnes qui ont toutes les apparences de la mort. Chambery, 1785.

[In 1784 the Imperial and Royal Academy of Sciences, etc., of Brussels proposed as a subject for a prize essay, What are the means that can be employed by medicine and police to prevent the dangerous mistakes of premature burial?]

WAUTERS (Pierre Englebert). Responsum ad quaesitum, Quae tum medica, tum politica praesidia adversus periculosas inhumationum praefestinatarum abusus? Reprinted from the Mem. Acad. Imper. et Roy. de Sc. de Bruxelles. Bruxelles, 1787 [1788].

PREVINAIRE (P. J. B.). Mémoire sur la question suivante proposée en 1784 par l'academie imperiale et royal des sciences, belles-lettres, et arts de Bruxelles : Quels sont les moyens que la médecine et la police pourroient employer pour prévenir les erreurs dangereuses des enter-remens precipités? Ouvrage qui a concouru pour la prix de l'annee 1786. Bruxelles, 1787.

———— The above in a German translation by Bernhard Gottlob Schreger. Leipzig, 1790.

LEDULX (Gul. Petrus). De signis mortis rite aestimandis. Hardervici, 1787. Thesis.

THIERY (Franciscus). La vie de l' homme respectée et defendue dans ses derniers moments; ou instruction sur les soins qu' on doit aux morts, et à ceux qui parroisent l' etre; sur les funerailles et les sepultures. Paris, 1787.

STEINFELD (Johannes Christianus). De signis mortis diagnosticis dubiis cautè admittendis et reprobandis. Thesis. Jena, 1788.

HERZ (Marcus). Ueber die frühe Beerdigung der Juden. Zweite vermehrte Auflage. Berlin, 1788.

DURANDE (J. Fr.). Mémoire sur l' abus de l' ensevelissement des morts, etc. Strasbourg, 1789.

DE HUPSCH (Baron Joh. Wilh. Carl Adolph). Nouvelle découverte d'une methode peu couteuse, efficace et assurée de traiter tous les hommes décedés afin de rappeler à la vie ceux qui ne sont morts qu'en apparance. Cologne, 1789.

ANON. Des inhumations precipitées. Paris, 1790. (Attributed by Barbier to Madame Necker.)

HUFELAND (Christoph W.). Ueber die Ungewissheit des Todes, und des einzige untrügliche Mittel . . . das Lebenigbegraben unmöglich zu machen, etc. Salzburg, 1791; Halle, 1824.

REINHARDT (Julius Christophorus). Dissertatio de vano praematurae sepulturae metu. Jena, 1793.

MARCELLO (Marin). Osservazioni teoriche-pratiche-mediche sopra le morti apparenti. Two vols., with nine plates. Venezia, 1793.

ANSCHEL (Salomon). Thanatologia, sive in mortis naturam causas genera, etc., disquisitiones. Goettingae, 1795.

HIMLY (Carolus). Commentatio mortis historiam causas et signa sistens. Goettingae, 1795.

PESSLER (B. G.). Leicht anwendbarer Beystand der Mechanik um Schein-todte beim Erwachen im Grabe auf die wohlfeilste Art wieder daraus zu erretten. Braunschweig, 1798.

DESESSARTZ (Jean Charles). Discours sur les inhumations precipitées. Paris, an vii. (1798).

KÖPPEN (Heinrich Friedrich). Nachrichten von Menschen welche lebendig begraben worden. Als erster Theil des Buchs : Achtung der Scheintodten. Halle, 1799. (Dedication to Friedrich Wilhelm III., King of Prussia, Queen Louise, and Friedrich August, Prince of Hesse Darmstadt.)

RESUSCITATION OF THE DROWNED.—THE ROYAL HUMANE SOCIETY.

GRUNER (Jacobus). Dissertatio inauguralis de causa mortis submer-sorum eorumque resuscitatione observationibus indagata. Groningae, 1761.

Memoirs of the society instituted at Amsterdam in favour of drowned persons. For the years 1767-71. Translated by Thomas Logan, M.D. London, 1772.

JOHNSON (Alexander), M.D. A short account of a society in Amster-dam . . . for the recovery of drowned persons ; with observa-tions showing the advantage . . . to Great Britain from a similar institution . . . Extended to other accidents. London, 1773.

JOHNSON (Alexander), M.D. A collection of cases proving the practicability of recovering persons visibly dead, etc. London, 1773.

———— Relief from accidental death ; or, summary instructions for the general institution proposed in 1773. London, 1785.

———— Abridged instructions. London, 1785.

CULLEN (W.), M.D. A letter to Lord Cathcart concerning the recovery of the drowned and seemingly dead. London, 1773.

HUNTER (John). Proposals for the recovery of persons apparently drowned. *Phil. Trans.* 1776.

HAWES (William), M.D. An address to the public [concerning the dangerous custom of laying out persons as soon as respiration ceases]. With a reply by W. Renwick, and observations on that reply. London, 1778.

FULLER (John), M.D. Some hints relative to the recovery of persons drowned and apparently dead. London, 1784.

KITE (Charles), of Gravesend. An essay on the recovery of the apparently dead. London, 1788.

———— Essay on the submersions of animals. London, 1795.

Reports of the Humane Society for the recovery of persons apparently drowned. For the years 1777-80 and 1785-86. London.

The transactions of the Royal Humane Society from 1774 to 1784. With an appendix of miscellaneous observations on suspended animation. Edited by W. Hawes, M.D. London, 1794.

FRANKS (John). Observations on animal life and apparent death. With remarks on the Brunonian system of medicine. London, 1790.

———— The same in an Italian translation. Pavia, 1795.

GOODWYN (Edmund), M.D. De morbo morteque submersorum investigandis. Thesis. Edin., 1786.

———— The connexion of life with respiration ; or, an experimental inquiry into the effects of submersion, strangulation, and several kinds of noxious airs on living animals ; with an account of the nature of the diseases they produce, and the most effectual means of cure. London, 1788.

Reflections on premature death and premature interment. Published by the Humane Society. Rochester, 1787.

ANON. An essay on vital suspension : being an attempt to investigate and ascertain those diseases in which the principles of life are apparently extinguished. By a Medical Practitioner. London, 1791.

HAMILTON (Robert), M.D. Rules for recovering persons recently drowned. London, 1795.

Directions for recovering persons apparently dead from drowning, and from disorders occasioned by cold liquors. Published by the Humane Society. Philadelphia.

CURRY (James). Popular observations on apparent death from drowning, suffocation, etc. Northampton, 1792 ; London, 1793, 1797, 1845. French transl. by Odier, Geneva, 1800.

FOTHERGILL (Anthony). Inquiry into the suspension of vital action in drowning and suffocation. Third ed. Bath, 1794.

————— Preventive plan ; or, hints for the preservation of persons exposed to accidents which suspend vital action. London, 1798.

CAILLEAU (J. M.). Mémoire sur l'asphyxie par submersion. Bordeaux, 1799.

BICHAT (M. F. Xavier). Recherches physiologiques sur la vie et la mort. Paris, 1800, 1805, etc.

NINETEENTH CENTURY.

COLEMAN (Edward). Dissertation on natural and suspended respiration. Second ed. Lond., 1802.

STRUVE (Christian August). A practical essay on the art of recovering suspended animation. Transl. from the German. Second ed. Lond., 1802.

OSWALD (John). On the phenomena of suspended animation from drowning, hanging, etc., together with the most expeditious mode of treatment. Philad., 1802.

LUGA (————). Traitement des asphyxiés, ou moyen de rendre impossible l' enterrement de personnes vivantes. Paris, 1804.

ACKERMANN (J. F.). Der Scheintod und das Rettungsverfahren. Frankft., 1804.

BURKE (William). On suspended animation, etc. Lond., 1805.

BERGER (J. F.). Essai physiologique sur la cause de l'asphyxie par submersion. Paris, 1805.

THOMASSIN (J. Franç.). Considerations de police médicale, sur la mort apparente, et sur le danger des inhumations precipitées. Strasbourg, 1805. Also an earlier essay on same subject, with Durande, in 1789.

DAVIS (——). L'abus des enterrements précipitées. Moyens de rappeler à la vie les personnes en état de mort apparente. Verdun, 1806.

BARZELOTTI (Giac.). Memoria per servire di avviso al populo sulle asfisse o morte apparente. Parma, 1808.

MARC (C. C. H.). Des moyens de constater la mort par submersion. (Manuel de l'Autopsie, par Rose, transl. from the German.) Paris, 1808.

COLORINI (Ant.). Sulle varie morti apparenti, etc. Pavia, 1813.

PORTAL (A.). Sur la traitement des asphyxies : avec observations sur les signes qui distinguent la mort réelle de celle qui n'est qu' apparante. Paris, 1816.

ORFILA (F.). Directions for the treatment of persons who have taken poison, and those in a state of apparent death. Transl. from the French by R. H. Black. Other transl. by W. Price, M.D. Both at London, 1818.

SNART (John). Thesaurus of horror ; or, the charnel-house explored, Lond., 1817.

——— An historical inquiry concerning apparent death and premature interment. London, 1824.

VALPY (R.). Sermon before the Royal Humane Society, with observations on resuscitation. Norwich, 1819.

WHITER (Rev. W.). A dissertation on the disorder called suspended animation. Norwich, 1819.

CHAUSSIER (——). Vivants crus morts, et moyens de prévenir cette erreur. Paris, 1819.

DONNDORF (J. A.). Ueber Tod, Scheintod, und zu frühe Beerdigung. Quedlinburg, 1820.

HERPIN (M.). Instruction sur les soins à donner aux personnes asphyxiées. Paris, 1822.

KAISER (Ch. L.). Ueber Tod und Scheintod, oder die Gefahren des frühen Begrabens. Frankfurt-am-Main, 1822.

CALHOUN (T.). An essay on suspended animation. Philad., 1823.

BUNOUST (Marin). Vues philanthropiques sur l'abus des enterrements précipitées, précautions à prendre pour que les vivants ne soient pas confondus avec les morts. Arras, 1826.

SPEYER (Carl F.) Ueber die Möglichkeit des Lebendigbegrabens, und die Einrichtung von Leichenhäusern. Erlangen, 1826.

CHANTOURELLE (——). Paper at the Royal Academy of Medicine of Paris, on the danger of premature burial, etc., with discussion thereon, 10th and 27th April, 1827. Archives générales de médecine, vol. xiv. (1827), p. 103.

GÜNTHER (Johann Arnold). Geschichte und Einrichtung der Hamburgischen Rettungs-Anstalten für im Wasser verunglückte Menschen. Hamburg, 1828.

TABERGER (Joh. Gottt.). Der Scheintod in seinen Beziehungen auf das Erwachen in Grabe und die verchiedenen Vorschläge zu einer wirksamen . . . Rettung in Fällen dieser Art. With a copper plate. Hannover, 1829.

BOURGEOIS (R.). Observations et considérations pratiques qui établissent la possibilité du retour à la vie dans plusieurs cas d'asphyxié et de syncope prolongée avec apparence de la mort. 8vo. Paris, 1829.

SCHNEIDAWIND (Franz Joseph Adolph). Der Scheintod, nebst Unterscheidung des scheinbaren und wahren Todes, und Mitteln, etc. Bamberg, 1829.

WALKER (G. A.). Gatherings from grave-yards, etc. Lond., 1830.

TACHERON. De la vérification légale des décès dans la ville de Paris, et de la nécessité d' apporter dans ce service médical plus de surveillance. Paris, 1830.

PICHARD (——). Le danger des inhumations précipitées. Paris, 1830.

CHAUSSIER (Hector). Histoire des infortunés qui ont été enterrés vivants. Paris, 1833.

DESBERGER (Ant. F. A.). Tod und Scheintod, Leichen-und-Begrabungs-wesen als wichtige Angelegenheit der einzelnen Menschen und des Staates. Leipzig, 1833.

FOUCHARD (P.). Aperçu général des précautions prises en France avant l'inhumation des citoyens morts ; réforme que l'humanité réclame. Tours, 1833.

DE FONTENELLE (Julia). Recherches médico-legales sur l'incertitude des signes de la mort, les dangers des inhumations précipitées, les moyens de constater les décès et de rappeler à la vie ceux qui sont en état de mort apparente. Paris, 1834.

LEGALLOIS (C.). Expériences physiologiques sur les animaux tendant à faire connaitre le temps durant lequel ils peuvent être sans danger privés de la respiration, etc. Paris, 1835.

MARC (C. C. H.). Nouvelles recherches sur les recours à donner aux noyés et asphyxiés. Paris, 1835.

SOMMER (——). De signis mortem hominis absolutam ante putredinis accessum indicantibus. Havniae, 1833.

SCHWABE (C.). Das Leichenhaus in Weimar. Nebst einigen Worten über den Scheintod und mehrer, jetzt bestehender Leichenhäuser, sowie über die zweckmässigste Einrichtung solcher Anstalten im Allge-meinen. Leipzig, 1834.

KAY (J. P.). The physiology, pathology, and treatment of asphyxia, including suspended animation in new-born children, and from drowning, hanging, wounds of the chest, mechanical obstruction of the air-passages, respiration of gases, death from cold, etc. London, 1834.

KOOL (J. A.). Tabellarisch overzigt over alle gevallen von schijndoode drenkelingen, gestikten, en gehangenen, bekroond door de Maat-schappij tot Redding van Drenkelingen, opgerigt in den jare 1767 te Amsterdam. Sedert hare stichting tot en met den jare 1833 [-53]. Uit authentieke stukken opgemaakt en met opmerkingen voorzien. Four vols. Amsterdam, 1834-54.

MANNI (Pietro), professor at Rome. Manuale pratico per la cura degli apparentemente morti, premessevi alcune idee generali di polizia medica per la tutela della vita degli asfittici. Roma, 1833. Napoli, 1835. Germ. transl. by A. F. Fischer, Leipzig, 1839.

SIMON (L. C.). Quelques mots sur les enterrements prématures, et sur les précautions à prendre sur-le-champ, relativement aux noyés et asphyxiés. St. Petersbourg, 1835.

LE GUERN (H.). Rosoline, ou les mystères de la tombe. Paris, 1834.

———— Du danger des inhumations précipitées, exemples tant anciens que récents de personnes enterrées ou dissequées de leur vivant. Paris, 1837, 1844.

———— Encore un mot, etc. Paris, 1843.

LESSING (Mich. Bened.). Ueber die Unsicherheit der Erkenntniss des erloschenen Lebens, etc. Berlin, 1836.

SCHNACKENBERG (Wilh. Ph. J.). Ueber die Nothwendigkeit der Leichenhallen zur Verhütung des Erwachens im Grabe. Cassel, 1836.

MISSIRINI (Melchiore). Pericolo di seppillire gli uomini vivi creduti morti. Milano, 1837.

VIGNE (——). Memoire sur les inhumations précipitées, des moyens de les prevenir, des signes de la mort. Rouen, 1837 ; Paris, 1839, 1841.

BIOPHILOS. Die neue Sicherungsweise gegen rettungloses Wiedererwachen im Grabe. Neustadt, 1838.

SCHAFFER (Fried.). Beschreibung und Abbildung einer Vorrichtung durch welche Scheintodte sich aus dem Sarge in Grabe befreien können. Landsberg, 1839.

VILLENEUVE (P. E.). Du danger des inhumations précipitées et des moyens de les prévenir, etc. Paris, 1841.

DESCHAMPS (M. H.) Précis de la mort apparente. Paris, 1841.

———— Du signe de la mort réelle, etc Memoir read at the Acad. des Sc., March 28, 1843, in Gaz. Med., Ap. 1st.

———— Du signe certain de la mort, nouvelle epreuve pour éviter d'etre enterré vivant. Paris, 1854.

NASSE (Fried.). Die Unterscheidung des Scheintodes von wirklichen Tode, zu Beruhigung über die Gefahr lebendig begraben zu werden. Bonn, 1841. French transl. by Fallot. Namur, 1842.

HICKMANN (J. N.). Die Elektricität als Prüfungs-und-Belebungsmittel im Scheintode. Wien, 1841.

25

DENDY (W. C.).   The philosophy of mystery, etc.   London, 1841.
    [Contains chapters on premature interment, resuscitation from cata-
    lepsy or trance, etc.]

WELCHMAN (E.).   Observations on apparent death from suffocation or
    drowning, choke-damp, stroke of lightning, exposure to extreme cold,
    with directions for using the resuscitating apparatus invented by author,
    and gen. instruc., etc.   8vo.   New York, 1842.

LENORMAND (Leonce).   Des inhumations précipitées.   Macon, 1843.

GAYET (——).   De la nécessité de la verification des décès   Nantes, 1843.

CHALETTE (J.), fils.   Du danger des inhumations précipitées et de
    l'importance de faire constater les décès par les gens de l'art.   Chalons-
    sur-Marne, 1843.

BARJAVEL (C. F. H.).   Nécessité absolue d'ouvrir au plus tôt des maisons
    d'attente ; considérations de police médicale, précedées d'un som-
    maire analytique, et suivies d'indications bibliographiques relatives
    au sujet de cet écrit.   (Tirage à cinquante exemplaires seulement).
    Carpentras, 1845.

DEBAY (Auguste).   Les vivants enterrés et les morts resuscités.   Con-
    siderations physiologiques sur les morts apparentes et les inhumations
    précipitées.   Paris, 1846.

GAILLARD (X.).   Préservatif contre le danger d'être enterré vivant, ou
    devoirs sacrés des vivants envers les morts.   Paris, 1847.

LOTHMAR (C. J.).   Ueber das Lebendigbegraben.   Leipzig, 1847.

DU FAY (Hortense G.).   Des vols d'enfant, et des inhumations d'individus
    vivants, suivi d'un aperçu pour l'etablissement des salles mortuaires.
    Paris, 1847.

    [In 1839 the Paris Academie des Sciences threw open to competition the
    Prix Manni (1,500 francs, founded in 1837 by Professor Manni, of Rome,)
    for the best work on the signs of death and the means of preventing pre-
    mature burials.   The prize was not assigned on that occasion, nor in 1842;
    but in the competition of 1846 it was assigned to Bouchut, on the report to
    the Academy by Rayer, May 29, 1848.]

BOUCHUT (E.).   Traité des signes de la mort et des moyens de ne pas être
    enterré vivant.   Paris, 1849.   Second ed., 1847 ; third ed., 1883.

——————   Mémoire sur plusieurs nouveaux signes de la mort, fournis par
    l'opthalmoscopie, et pouvant empecher les enterrements précipitées.
    Paris, 1867.

BRAID (James). Observations on trance, or human hybernation. London, 1850.

KAUFMANN (M.). De la mort apparente et des enterrements précipités. Paris, 1851.

KERTHOMAS (Hyac. L. De). Inhumations précipitées. Lille, 1852.

HARRISON (James Bower). The medical aspects of death. Lond., 1852.

CRIMOTEL (J. B. Valentin). Des inhumations précipitées; épreuve infaillible pour constater la mort; moyens de rappeler à la vie dans les cas de mort apparente causée par l'ether, le chloroforme, etc. Paris, 1852.

———— De l'épreuve galvanique ou bioscopie électrique, moyens de reconnaître la vie ou la mort et d'eviter les inhumations précipitées. 1866.

JOSAT (———). De la mort et ses caractères. Necessité d' une révision de la législation des décès pour prevenir les inhumations et les délaissements anticipés. Ouvrage entrepris et exécuté sous les auspices du gouvernement et couronné par l' Institut. Paris, 1854.

LONDE (C.). Lettre sur la mort apparente, les conséquences réelles des inhumations précipitées, le temps pendant lequel peut persister l'aptitude à être rappelé à la vie. Paris, 1854. Plates.

KEMPNER (F.). Denkschrift über die Nothwendigkeit einer gesetzlichen Einführung von Leichenhäusern. New ed. Breslau, 1856.

PEYRIER (J. P. P.). Récherches sur l'incertitude des signes de la mort : enumeration des maladies qui peuvent produire la mort apparente ; abus des enterrements précipités. Paris, 1855.

COLLONGUES (L.). Application de la dynamoscopie à la constatation des décès. Paris, 1858, 1862.

HALMA GRAND (———). Des inhumations précipitées. Paris, 1860.

WELBY (Horace). Mysteries of life, death, and futurity (with chapter on premature interment). London, 1861.

REYHER (O. C. A.). Ueber die Verwerthung der bekannten Leichenerscheinungen zur Constatirung des wahren Todes. Leipzig, 1862.

CHEVANDIERE (Antoine Daniel). De la vérification des décès et de l'organisation de la medecine cantonale. Paris, 1862.

DESMAIRE (Paul). Les morts vivants. Paris, 1862.

BARRANGEARD (Antoine). Extrait de divers mémoires publies depuis tres longtemps par le Docteur Barrangeard, sur le danger des inhumations précipitées et sur l'indispensable nécessité de constater avec soin tous les décès sans exception. Lyon, 1863.

BONNEJOY (E.). Des moyens pratiques de constater la mort par l'électricité à la aide de la faradisation. Paris, 1866.

LEVASSEUR (P.). De la catalepsie au point de vue du diagnostic de la mort apparente. 8vo. Rouen, 1866.

———— De la mort apparente et des moyens de la reconnaître. Rouen, 1867. Re-issued, with a second essay, in 1870.

JACQUAND (Frédéric). Appareil respiratoire avertisseur pour les tombes. Assurance contre la mort apparente. Paris, 1867.

BIANCO (Giuseppe). Le pericolose consequenze della morte apparente prevenute da un confaciente riforma del servizio mortuario. Torino, 1868.

GANNAL (Félix). Mort apparente et mort réelle. Moyens de les distinguer. First ed. Paris, 1868. Third ed. (mention honorable á l'Institut de France), 1890.

[In 1868 the Académie de Médecine of Paris threw open to competition the Prix d' Ourches of 20,000 francs for the discovery of a simple and popular means of detecting the signs of real death certainly and beyond doubt. The prize was not awarded, but premiums were given to several competitors.]

HOARAU (H.). La mort, sa constatation, ou procédé à l'aide du quel on peut la reconnaître et éviter des enterrements de vifs. Paris, 1874.

VEYNE (——). Mort apparente et mort réelle, artériotomie donnant le moyen de les reconnaître. Paris, 1874.

MONTEVERDI (A.). Note sur un moyen simple, facile, prompt et certain de distinguer la mort vrai de la mort apparente de l'homme. Cremone, 1874.

MARTEL (——). La mort apparente chez les nouveaux-nés. Paris, 1874.

BOILLET (Ch.). Mort apparente et victimes ignorées. Paris, 1875.

DE COMEAU (——). Les signes certains de la mort mis à la portée de tout le monde. Limoges, 1876.

BELVAL (Th.). Les maisons mortuaires. Paris, 1877.

FRITZ-ANDRE (——). Du danger des inhumations précipitées. Bruxelles, 1879.

[The Prix Dusgate was founded by a decree of November 27, 1874, authorising the Académie des Sciences of Paris to accept the legacy of M. Dusgate of a quinquennial prize of 2,500 francs for the best work on the diagnostic signs of death and on the means of preventing premature burial. The essays of the first competition were received on June 1, 1880, and on March 14, 1881, the prize was divided among the three following competitors. In 1885 the prize was not awarded.]

ONIMUS (E. N. J.). Modification de l'excitabilité des nerfs et des muscles apres la mort. (Published.)

PEYRAND (H.). De la détermination de la mort réelle par le caustique de Vienne.

LE BON (G.). Recherches experimentales sur les signes diagnostiques de la mort et sur les moyens de prevenir les inhumations précipitées. (A temperature of 25° C. on a thermometer kept in the mouth for a quarter of an hour.) Also, Article on Premature Interment in Monit. scient., viii. Paris.

ALLEN (F. D.). Remarks on the dangers and duties of sepulture, or security for the living with respect and repose of the dead. Boston, 1873.

BURDETT (H. C.). The necessity and importance of mortuaries for towns and villages, with suggestions for their establishment and management. London, 1880.

FLETCHER (Moore Russell). One thousand persons buried alive by their best friends. A treatise on suspended animation, with directions for restoration. Boston, 1890.

"A Hygienic Physician." Earth to earth burial and cremation by fire [includes cases of premature burial]. London, 1890.

HERNANDEZ (Maxime F. E. M.). Contribution à l'étude de la mort apparente. Bordeaux, 1893.

LIGNIERES (Dr. D. De). Ne pas être enterré vivant. Paris, 1893.

Traitement physiologique de la mort apparente. Series of twenty-five papers in "La Tribune Médicale," Paris, 1894, vol. xxvi., 2 ser.

GILES (Alfred E)). Funerals, suspended animation, premature burials, Boston, 1895.

GAUBERT (B.), Avocat. Les chambres mortuaires d'attente, devant l'histoire, la legislation, la science, l'hygiène et le culte des morts. (Le péril des inhumations précipitées en France.) With sixty figures, maps or plans. Paris, 1895.

HARTMANN (Franz). Buried alive: An examination into the occult causes of apparent death, trance, and catalepsy. Boston, U.S., 1895. Lond., 1896. Also, Lebendig begraben. Leipzig, 1896.

WILDER (Alexander). The perils of premature burial. London, 1895.

French theses (at Paris, unless otherwise stated,) on apparent death, the signs of death, danger of premature burial, etc. :—

JOUY (Montpellier), 1803.
THOMASSIN (Strassbourg), 1805.
LAURENT, 1805.
PIERRET, 1807.
VERNEY, 1811.
FOUCHER, 1817.
GRESLON, 1819.
FERRY, 1819.
LEPAULMIER, 1819.
LEVY (Strassbourg), 1820.
AMAND D'AMBRAINE, 1821.
POUIER, 1823.
WEST, 1827.
PIERRET, 1827.
GLEIZAL, 1829.

D'ALENCASTRE, 1832.
CHAMPNEUF, 1832.
BONIFACE, 1833.
LINARES, 1834.
MENESTREL, 1838.
DE SILVEIRA PINTO, 1837.
CARRE, 1845.
DOSAIS, 1858.
GRESLON, 1858.
PARROT, 1860.
LEGLUDIC, 1863.
SCHNEIDER (Strassbourg), 1863.
ACOSTA, 1864.
EDMOND, 1871.

Graduation theses other than French, on the same theme :—

VAN GEEST (Lugd. Bat.), 1811.
DAVIES (Edin.), 1813.
GOURY (Leodii), 1828.
TSCHERNER (Breslau), 1829.
SOMMER (Havniae), 1833.
NYMAN (Dorpat), 1835.

BETTMAN (Munich), 1839.
SCHMIDT (Nürnberg), 1841.
KLUGE (Leipzig), 1842.
WENDLER (Leipzig), 1845.
KRIBBEN (Bonn), 1873.
SORGENFREY (Dorpat), 1876.

FRENCH ARTICLES IN JOURNALS.

ABADIE (C.). Note sur l'examen ophthalmoscopique du fond de l'oeil comme signe de la mort réelle. Gaz. d'Hôp., vol. xlvii, p. 290. Par., 1874.

BOUCHUT (E.). Mort apparente durant six heures, avec absence des battements du coeur à l'auscultation. Gaz. d'Hôp., vol. xxvii., p. 223. Par., 1854.

BOURGEOIS (R.). Du danger d'être enterré vivant et des moyens de constater la mort. Bull. Acad. de Méd., vol. ii., pp. 619-626. Paris, 1837-38, and Rev. Méd. Franç. et étrang., vol. ii., pp. 360-378. Paris, 1838.

BROWN-SÉQUARD (——). "Extraordinary prolongation of the principal acts of life after the cessation of respiration." Arch. de Physiol. Norm. et Path., vol. vi., 2 S., pp. 83-88. Par., 1879.

—————— "Researches on the possibility of recalling temporarily to life persons dying of sickness." J. de la Physiol. de l'Homme, vol. i., pp. 666-672. Par., 1858.

CAZIN (——). De la nécessité de faire constater tous les genres de mort. Précis d'Trav. Soc. Méd. de Boulogne-sur-mer, vol. i., pp. 27-33. 1839.

CHAUSSIER (——). Rapport sur les enterremens précipités. Bull. Fac de Méd. de Par., vol. v., pp. 467-476. 1816-17.

DESCHAMPS (M.-H.). Mémoire sur la vérification des décès et sur le danger des déclarations précipitées. Union Med., vol. xxi., N.S., pp. 56, 106. Par., 1864.

DEVERGIE (——). Inhumations précipitées. Ann. d'Hyg., 2 S., vol. xxvii., pp. 293-327. Paris, 1867. De la création de maisons mortuaires et de la valeur des signes de la mort. Ann. d'Hyg., vol. xxxiv., 2 S., pp. 310-327. Par., 1870.

—————— Des signes de la mort; étude de leur cause, appréciation de leur valeur. Ann. d'Hyg., vol. xli., 2 S., pp. 380-405. Par., 1874.

FODERE (——). Signes de la mort. Dict. de Sc. Med., vol. li., pp. 294-306. Paris, 1821.

FOUANES (——). Sur la rigidité cadavérique comme signe certain de la mort. Gaz. Med. de Par., vol i., 3 S., p. 91. 1846.

FOUQUET (——). Mémoire sur la roideur cadavérique considerée comme signe certain de la mort. Gaz. Med. de Par., vol. ii., 3 S., pp. 250-255. 1847.

FOURNIE (——). Les signes de la mort et le prix d'Ourches. (Also translated into Italian.) Gaz. d'Hôp., vol. xlvii., pp. 273-275. Par., 1874.

GIRBAL (——). Mort apparente: mesures prématurés d'inhumation: topiques stimulants, prompte cessation des phénomenes léthiformes, guérison. Revue de Thérap. du midi, vol. ii., pp. 161-167. Montpellier, 1851. Also, Gaz. d'Hôp., vol. iii., 3 S., p. 142. Par., 1851.

GRETSCHER DE WANDELBURG. (For Marquis d'Ourches's prize.) Des moyens de distinguer la mort réelle de la mort apparente. In his Mém. de Méd. et Chir., pp. 49-54. 8vo. Par., 1881.

HAMON (L.). Simple note sur la mort apparente; acupuncture cardiaque et diaphragmatique. Rev. de Thérap Med. Chir., vol. xlvii., p. 482. Par., 1880.

HENROT (H.). Persistance des battements du cœur pendant plus d'une heure après la cessation de la respiration. Bull. Soc. Méd. de Reims., No. 15, pp. 139-144. 1876-77.

LABORDE (J. V.). Gaz. hebd. de. Méd., vol. viii., 2 S., pp. 605, 623, 710. Par., 1871.

LARCHER (——). Arch. gén. de Méd., vol. i., pp. 685-709. Par., 1862.

LEGRAND (A.). Rev. Méd. Franç. et étrang., vol. i., pp. 705-714. Par., 1850.

LEVASSEUR (P.) et MARTINS (S.). France Méd., vol. xiv., pp. 169, 177, 204, 226, 228. Par., 1867.

MALHOL (J.). Journ. Gén. de Méd. Chir. et Pharm., vol. xxii., p. 470. Par., 1805.

MICHEL (A.). Bull. gén. de Therap., etc., vol. xxxvii., pp. 462-464. Par., 1849.

MONFALCON (J. B.). Art. "Mort," Dict. de Sc. Méd., vol. xxxiv., pp. 319-347. Par., 1819.

NICATI (W.). Un signe de mort certaine, emprunti à l'ophthalmotonométrie; lois de la tension oculaire. Compt. Rend. Acad. de Sc. cxviii., p. 206. Paris, 1896.

PAPILLON (F.). Rev. des Deux Mondes, vol. civ.. pp. 669-688. Par., 1873.

PINGAULT (——). Bull. Soc. de Méd. de Poitiers. vol. xxviii., pp. 83-86. 1860.

PLOUVIEZ (——). Union Méd. Paris, vol. i., pp. 408-424. 1870.

Report to French Academy of Sciences on apparent deaths, etc., by Rayer. Compt. Rend. Acad. de Sc. (Séance, May 29, 1848.) Also in Ann. d'Hyg.. vol. xl., pp. 78-110. Par., 1848; and in Ann. de Méd. Belge., vol. lv., pp. 1-24. Brux., 1848; and in Bull. Soc. de Méd. de Poitiers, vol. xv., pp. 39-53, 1849.

SIMON (A.). Bull. gén. de Therap., etc., vol. xxxvii.. pp. 221-226. Par., 1849.

SIMONOT (——). Union Méd. de Par., vol. xii., 2 S., pp. 211. 286. 1862.

TOURDES G.). Art. "Mort : la mort apparente," in Dict. Encycl. d. Sc. Méd., vol. ix., 2 S., pp. 598-690. Par., 1875.

TOURNIE (——). Union Méd., vol. viii., p. 235. Par., 1854.

VAN GHEEL (——). Gaz. d. Hôp., vol. xliv., pp. 345, 353. Par.. 1871.

VAN HENGEL (J.). Journ. de Méd. Chir. et Pharm. Col., vol. vi., pp. 523-525. Brux., 1848.

GERMAN ARTICLES. (The Titles Translated.)

ALKEN (——). Restoration to life of one apparently dead. Wochenschr. f. d. ges. Heilk., p. 319. Berlin, 1838.

ARNOLD (J. W.). On acupuncture of the heart as a means of recovery in apparent death. Heidlb. klin. Ann., vol. vii., p. 311. 1831.

BALDINGER (E. G.). Literary contribution to the history of being buried alive. N. Magaz. f. Aerzte., vol. xiv., p. 84. Leipzig, 1792.

BETZ (F.). Sudden apparent death in a child with vomiting and purging. Memorab., vol. v., p. 119. Heilbrn., 1860.

DEUBEL (——). New and simple means for the recovery of the apparently dead. Wochenschr. f. d. ges. Heilk., p. 597. Berlin, 1846.

DIRUF (——). On the dread of being buried alive, etc. Ztschr. f. d. Staatsarznk., extra part. p. 72. Erlang., 1840.

DYES (A.). Apparent death caused by inflammation of the lungs. Deutsche Klinik, vol. xxiii., p. 44. Berl., 1871.

HANDSCHUH (——). A few remarks on mortuaries as a means of preventing the burial of the apparently dead. Ztschr. f. d. Staatsarznk., vol. xxi., p. 34. Erlang., 1831.

HECHT (S. C.). Reflections and proposals concerning the impracticability of the existing regulations to prevent the burial of the apparently dead. Ann. d. Staatsarznk., vol. v., p. 395. Freib., 1840.

HOFFMANN (—). Simple means of preventing the being buried alive. Allg. Med. Centr. Ztg., vol. xvi., p. 609. Berl., 1847.

HOPPE (J.). Recovery of one apparently dead and of one dying, by burning on the breast. Memorabilien, vol. vi., p. 199. Heilbrn., 1861.

HUBER (M.). On inspection of the dead. Ztschr. d. Gesellsch d. Aerzte zu Wien, vol. ii., p. 120. 1853.

HUFELAND (——). Report on the certain and uncertain signs of death, on the indications of returning vitality, and how one should deal with corpses in general. Weimar ordinance, 1794. Beytr. z. Arch. d. Med. pol., vol. vii., I S., p. 61. Leipzig, 1797.

KAISER (K. L.). What means has the State to take so as to ensure that no one be buried alive? Ztschr. f. d. Staatsarznk., fourteenth extra number, p. 100. Erlang., 1831.

KLEIN (F. X.). Metallic irritation as a means of proving death. Extract from Dissertation in Beytr. z. Arch. d. Med. pol., vol. vi., I S., p. 118. Leipzig, 1795.

KLOSE (C. L.). On the risk of being buried alive: several precautions against it. Ztschr. f. d. Staatsarznk., vol. xix., p. 143. Erlang., 1830.

KUNDE (F. T.). Physiological observations on apparent death. Arch. f. Anat. Physiol. u. wissenssch. Med., p. 280. Berlin, 1857.

MAGNUS (H.). Certificates of death and sanitary reports. Wochenschr. f. d. ges. Hlkde., p. 385. Berlin, 1841.

——— A certain sign that death has taken place. Virchow's Archiv., vol. lv., pp. 511, 523. 1872.

MASCHKA (J.). On symptoms of the corpse. Vrtljschr. f. d. prakt. Heilk., vol. iii., p. 91. Prag., 1851.

MASCHKA (J.). On diagnostic errors in medical jurisprudence. Vrtljschr. f. d. prakt. Heilk., vol. lxxix., p. 13. Prag., 1863.

MEYN (——). Fortunate resuscitation of an apparently dead woman. Mitth. a. d. Geb. d. Med. vi., Hft. 6-7, p. 76. Altona, 1838-9.

MOSSE (——). Certificates of death and sanitary reports. Wochenschr. f. d. ges. Heilk., p. 696. Berlin, 1842.

NASSE (F.). Measuring the temperature for the diagnosis of death. J. d. pract. Heilk., vol. xciii., 4 St., p. 130. Berl., 1841.

— Discrimination of apparent death from real death, to reassure as to the danger of being buried alive. Rev. of his essay (Bonn, 1841) in Mitth. a. d. Geb. d. Med., vol. ix., p. 11. Altona, 1841-43.

Ordinance of the Elector of Saxony concerning the treatment of corpses, and to provide against the premature interment of the apparent dead. Med. Chir. Ztg., vol. ii., p. 150. Salzburg, 1793.

PLAGGE (T.). Is the failure of the heart-beat a certain sign of death? Memorabilien, vol. v., p. 71. Heilbrn., 1860.

RADIUS (——). The awakening apparatus in the Leipzig Mortuary. Beitr. z. Prakt. Heilk., vol. i., p. 532. Leipzig, 1834.

RAMPOLD (——). On the inaudibility of the heart-beat as a sign of death. Cor. Bl. d. Württemb. ärztl. Vereins, vol. xxi., p. 353. Stuttg., 1851.

RÖSER (——). On being buried alive, and the mortuaries. Cor. Bl. d. Württemb. ärztl. Vereins, vol. xxvii., p. 115. Stuttg., 1857.

ROSENTHAL (M.) Researches and observations on the dying of the muscles, and on apparent death. Wien. med. Presse, vol. xiii., pp. 401, 419. 1872.

——— On the newest and safest means of knowing apparent death. Wien. med. Presse, vol. xvii., p. 461. 1876.

SCHMIDT (J. H.). On mortuaries, with a case of apparent death that did not end in death till twenty days after. Wochenschr. f. d. ges. Heilk., vol. i., p. 385. Berl., 1833.

SCHNEIDER (——). On the risk of being buried alive Ztschr. f. d. Staatsarznk., vol. xxxiv., p. 157. Erlang., 1837.

SICKLER (J. V.). Directions for preventing the burying of each other alive. Beytr. z. Arch. d. Med., 2 Samml., vol. iv., p. 158. Leipzig, 1793.

SPEYER (——). On the possibility of being buried alive, and on the erection of mortuaries. Ztschr. f. d. Staatsarznk., fifth extra part, p. 326. Erlang., 1826.

STRUVE (——). Simplified application of galvanism, etc., in cramps and in apparent death, and for proving actual death. J. d. Prakt. Arznk., 2 R., vol. xxiii., 4 St., p. 5. Berl., 1806.

TENGLER (G.). Critical remarks on the signs of death, with reference to the inspection of the dead. Wien. med. Wochenschr., vol. vii., p. 519. 1857.

THIERFELDER (——), sen. On apparent death and medical inspection of the dead. Deutsche Ztschr. f. d. Staatsarznk., vol. xxv. p. 241. Erlang., 1867.

VARGES (L.). On the awaking of one apparently dead. Ztschr. d. nordd. chir. Ver., vol. i., p. 353. Magdeb., 1847.

VON JÄGER (——). Account of an alleged coming to life in the grave. Ztschr. f. d. Staatsarznk, vol. vi., pp. 241-252. Erlang., 1823.

WILDBERG (C. F. L.). State precautions to obviate all anxiety as to being buried alive. Jahrb. d. ges. Staatsarznk, vol. iv., p. 169. Leipzig, 1838.

ZAUBZER (O.). Fragments on thanatology, for the police of the dead in Munich. Aerztl. Intellig. Bl., vol. xx., p. 106. München, 1874.

ENGLISH AND AMERICAN ARTICLES.

ALDIS (C. J. B.). On the danger of tying up the lower jaw immediately after supposed death. Lancet, vol. ii., 1850, p. 601.

ANON. Cases of apparent death. Calcutta J. M., vol. ii., pp. 380-387. 1869. From All the Year Round, July, 1869.

ANON. Signs of death. London M. Rec., vol. ii., pp. 205, 221. 1874.

BOURKE (M. W.). Resuscitation of a child after ten minutes' total submersion in water, etc. Dublin M. Press, vol. xliii., p. 103. 1859.

BRANDON (R.) Construction of houses for the reception of the dead; means for the recovery of those, etc. Med. Times, vol. xvi., p. 574. Lond., 1847.

CLARK (T. E.). Buried alive. Quart. Journ. Psych. Med., vol. v., pp. 87-93. N.Y., 1871.

COLDSTREAM (John). A case of catalepsy. Edin. Med. and Surg. Journ., vol. lxxxi., p. 477.

DANA (C. L.). The physiology of the phenomena of trance. Med. Rec., vol. xx., pp. 85-89. N.Y., 1881.

DAVIS (M.). Hasty burials. Sanit. Rec., vol. iv., p. 261. Lond., 1876.

DENMAN (J.). Resuscitation after two hours' apparent death (drowning). Med. Press. and Circ., vol. iii., p. 95. Dublin, 1867.

DOUGLAS (H. G.). Recovery after fourteen minutes' submersion. Lond. Med. Gaz., vol. i., p. 448. 1842.

DUCACHET (H. W.). On the signs of death, and the manner for distinguishing real from apparent death. Am. M. Recorder, vol. v., pp. 39-53. Phila., 1822.

FRASER (W.). Distinctions between real and apparent death. Pop. Sci. Month.. vol. xviii., pp., 401-408. New York, 1880-81.

GAIRDNER (W. T.). Case of lethargic stupor or trance, extending continuously over more than twenty-three weeks, etc. Lancet, vol. ii.. 1883. p. 1078, and vol. i., 1884, pp. 5, 56.

GOADBY (H.). Death trance. Med. Indep., vol. i., pp. 90-99. Detroit, 1856.

GODFREY (E. L. B.). Report of the resuscitation of a young girl apparently dead from drowning. Phila. M. Times, vol. ix., p. 375. 1879.

HUFFY (T. S.). Two cases of apparent death. Tr. M. Soc., N. Car., vol. xxi., pp. 126-131. Raleigh, 1874.

JAMIESON (W. A.). On a case of trance. Edin. Med. J., vol. xvii., pp. 29-31. 1871-72.

LEE (W.). The extreme rarity of premature burial. Pop. Sc. Month., vol. xvii., p. 526. N.Y., 1880.

MACKAY (G. E.). Premature burials. Ibid., vol. xvi., p. 389.

MADDEN (T. Moore). On lethargy or trance. Dubl. J. Med. Sc., vol. lxxi., p. 297. 1881.

MILLER (T. C.). The state of the eyelids after death—open or shut? Med. Rec., vol. xii., p. 4. N. Y., 1877.

OSBORNE (W. G.). Impositions of the Indian faqueer, who professed to be buried alive and resuscitated in ten months. Lancet, vol. i., 1839-40, p. 885.

POPE (C.). A case of recovery after long immersion. Lancet, vol. ii., 1881, p. 606.

POVALL (R.). An account of successful resuscitation of three persons from suspended animation by submersion for twenty-five minutes. West Med. and Phys. J., vol. ii., pp. 499-503. Cincin., 1828-29.

REID (T. J.). A case of suspended animation. St. Louis Clin. Rec., vol vi., pp. 261-263. 1879-80.

Report of Committee on suspended animation. Proc. Roy. M. and Chir. Soc. Lond., vol. iv. (1862), pp. 142-147 ; vol. vi. (1870), p. 299. See also Transactions, vol. xlv. (1862), p. 449.

RICHARDSON (B. W.). Researches on treatment of suspended animation. Brit. and For. M. Chir. Rev., vol. xxxi., pp. 478-505. London, 1863.

RICHARDSON (B. W.). The absolute signs and proofs of death. Asclepiad, No. 21. 1889.

ROMERO (Francisco). Infallible sign of extinction of vitality in sudden death. (Latin.) Med. Tr. Roy. Coll. Phys., vol. v., pp. 478-485. London, 1815.

SHROCK (N. M.). On the signs that distinguish real from apparent death. Transylv. J. M., vol. viii., pp. 210-220. Lexington, Ky., 1835.

SILVESTER (H. R.). A new method of resuscitating still-born children, and of restoring persons apparently drowned or dead. Brit. M. J., pp. 576-579. London, 1858.

TWEDELL (H. M.). Account of a man who submitted to be buried alive for a month at Jaisulmer, and was dug out alive at the expiration of that period. India J. M. and Phys. Sc., vol. i., N. S., pp. 389-391. Calcutta, 1836.

THOMAS (R. R. G.). The Marshall Hall method successful in a case of drowning of ten minutes' duration, and an interval of half an hour before its application. Lancet, vol. ii., 1857, p. 153.

TAYLOR (J.). Case of recovery from hanging. Glasg. Med. J., vol. xiv., p. 387. 1880.

WHITE (W. H.). A case of trance. Brit. M. J., vol. ii., 1884, page 52.

SPANISH ARTICLES.

ALCANTARA (F. C.). Enciel. Méd. Farm., vol. ii., pp. 265, 273, 275, 289, 297. Barcelona, 1878.

DEL VALLE (G.). An. r. Acad. de Cien. Méd. de la Habana. vol. viii., pp. 480-489. 1871-72.

GELABERT (E.). A case of premature interment. Rev. de Cien. Méd., vol. vii., pp. 67-69. Barcel., 1881.

GUEREJAZE (——). España Med., vol. x., p. 111. Madrid, 1865.

PULIDO (——). Anfiteatro Anat., vol. iv., pp. 164, 181. Madrid, 1876.

RAMON VIZCARRO. Siglo Méd., vol. xxvi., p. 777. Madrid, 1879.

—— Sentido Catól, vol. i., p. 284. Barcel., 1879.

ULLOA (——) Entierros prematuros. Gac. Méd. de Lima, vol. xii., p. 219. 1867-8.

ITALIAN ARTICLES.

BIANCO (G.). Report and discussion upon his work, "Dangers of Apparent Death" (Torino, 1868). Gior. d. r. Acad. di Med. di Torino., vol. vii., 3 S., pp. 243, 304, 366, 370. 1869.

CHIAPPELLI (G.). Sperimentale, vol. xliii., pp. 74-77. Firenze, 1879. Also in Gaz. Med. Ital. Prov. Venete, vol. xxii., p. 94. Padova, 1879.

IMPARATI (M.). Guglielmo da Saliceto, vol. ii., pp. 293, 325, 357. Piacenza, 1880-81.

PACINI (F.). Imparziale, vol. xvii., pp. 41, 75. Firenze, 1877.

PARI (A. D.). Arch. di Med. Chir. ed ig. Roma, vol. ix., p. 5-35. 1873.

SONSINO (P.). Imparziale, vol. vii., pp. 225-231. Firenze, 1867.

TAMASSIA (A.) and SCHLEMMER (A.). Riv. sper. di Freniat., vol. ii., pp. 628-639. Reggio-Emilia, 1876.

VERGA (A.) and BIFFI (S.). Gaz. Med. Ital. Lomb., vol. iii., 8 S., pp. 92-94. Milano, 1881.

ZILIOTTO (P.). Gior. Venete di Sc. Med., vol. i., 3 S., pp. 323-336. Venezia, 1864.

ZURADELLI(G.). Ann. univ. di Med., vol. vii., pp. 3-241. Milano, 1869.

INDEX.

26

*Hay Nisbet & Co., Printers, 16 St. Enoch Square, Glasgow, and
25 Bouverie Street, London, E.C.*